Field and Laboratory Exercises in Ecology

Field and Laboratory Exercises in Ecology

Stephen D. Wratten
Lecturer in Ecology, University of Southampton

Gary L. A. Fry
Assistant Regional Officer,
Nature Conservancy Council, Dalbeattie, Scotland

First published 1980
by Edward Arnold (Publishers) Limited
41 Bedford Square, London WC1B 3DQ

British Library Cataloguing in Publication Data

Wratten, Stephen
Field and laboratory exercises in ecology.
1. Ecology—Problems, exercises, etc.
I. Title II. Fry, Gary
574.5′076 QH541

ISBN 0-7131-2725-2

Printed in Great Britain by
Thomson Litho Ltd, East Kilbride, Scotland.

Preface

The aim of this book is to show how modern numerical techniques in plant and animal ecology can be used practically at school, college and undergraduate level to demonstrate many of the fundamental principles of the subject. Ecology is becoming increasingly quantitative but the application of this approach to practical classes is often restricted by timetables, availability of materials, size of classes and the need for lengthy preparation time. The exercises presented here are intended to show how practical ecology taught at this level need not suffer from the above restrictions and need not be confined to descriptive studies, while still fitting into a practical period of three hours or less.

Ecology is by definition a field subject but there are analytical elements which are well suited to laboratory investigation and simulation. In fact, most ecological investigations need a laboratory as well as a field component, as reference to modern journals will show. The arrangement of exercises in this book reflects this balance; they are presented in complementary field and laboratory pairs, each exercise of the pair using organisms and techniques according to the restrictions imposed by the experimental conditions and the aspect of the topic being investigated. For instance, in a pair of exercises on the dynamics of predation, the laboratory investigation concerns a simulation of the short-term behaviour of searching predators while the field exercise concerns the longer-term and larger-scale numerical response of predators to changes in the numbers of their prey. In the latter case, only a field approach provides the necessary realism.

As an introduction to each pair of exercises, the fundamental principles on which they are based are summarized. In each case this is followed by sections in which *apparatus*, *preparation* and *procedure* are described in detail for each exercise. A common *discussion* and *conclusions* section interprets the data collected and the reader is referred to published work to enable the results to be considered in a wider ecological context. This 'packaging' of exercises is a deliberate attempt to allow students to collect data from which meaningful ecological conclusions can be drawn in the limited time available. Such a guided discovery approach has been a frequent demand from teachers and lecturers in recent years. Many of the exercises incorporate modern ecological themes in a practical way for the first time at this level and a wide range of numerical techniques is employed. Through the inclusion of up-to-date introduction and conclusion sections the book is intended to be of use as a student manual rather than a teachers' guide.

Fifty-six exercises are described in detail, but in addition each pair is followed by a number of suggestions for related investigations so that the book contains almost two hundred workable practical ideas. Many of these are suitable for expansion into small research projects or for use on field courses when more time is available. With a little modification and extra preparation, many of the field exercises could be made into laboratory practicals.

The exercises are arranged in five subdivisions of the subject which obviate the often artificial segregation of plant and animal ecology: sampling, spatial pattern, populations, population interactions and community ecology. Each of these sections is preceded by a brief introduction which shows how the exercises relate to established work, but which for reasons of space is not an attempt at a full introduction to that ecological field.

Organisms used in these exercises all occur in Britain but related species in other temperate regions can easily be substituted without affecting the fundamentals of the methods. The choice of organisms and exercises is obviously biased towards our research and teaching interests, but in spite of this restriction the book includes exercises on fungi, algae, higher plants, crustaceans, insects, molluscs, birds and mammals. Wherever possible, procedures have been explained without the use of worked examples as these often give the impression that there is a 'correct' answer. Obviously the details of the results will change according to experimental conditions, so algebraic notation is used instead and the expected general trend of the results is indicated in the discussion. The appropriate statistical techniques are suggested for the analysis of results in each exercise and readers are referred to statistical textbooks (Bailey, 1959; Parker, 1979) for the detailed rationale of these tests. In those exercises in which invertebrates have to be identified to order or family, we recommend the illustrated keys in Lewis and Taylor (1967) and Chinery (1973). Field techniques and apparatus about which further information is required may be found in Chapman (1975) (plants) and Southwood (1971) (animals).

We are grateful to our colleagues Drs J. A. Allen and R. J. Putman who provided the ideas for exercises 41 and 43 (J. A. A.) and 55 (R. J. P.), to Mrs S. Meacock for helpful criticism of the manuscript, and to Mrs M. Lovell for doing most of the typing. Mrs J. Clayton agreed to our referring to an unpublished method. We would be pleased to receive comments on the success or otherwise of these exercises, which have given useful results when we have used them!

Southampton, Hampshire
1979

S. D. W
G. L. A. F.

Contents

Section One

Sampling

Introduction

Sampling is one of the most important yet often neglected parts of any ecological investigation. Since conclusions based on samples are used to make hypotheses about populations as a whole, the sampling procedure must be correct or the generalizations are invalid. Much attention is given to these aspects, for the study of plant populations by Greig-Smith (1964), and for animal populations by Southwood (1971). In this section the emphasis is on the general principles involved in sampling, and their application to a number of selected problems.

For the animal ecologist the problem of sampling is sometimes much reduced by the presence of natural sampling units. These are usually habitat units, for instance trees, dung, carrion, flowers or leaves. In contrast, most plant ecologists are forced to construct artificial sampling units. When this happens we have to decide where to place these units in space and what to record in them, how many are required to reflect the population accurately and sometimes how long the recording should progress.

Data collection procedures should aim to provide the maximum information within the experimental constraints of time, finance and manpower. Obviously no one method could be recommended for general usage since the effectiveness of any method depends on the study objectives and, in particular, the nature of subsequent analysis. However, two stages can be isolated for separate consideration; these are

1 *Sampling Strategies* A sampling strategy defines the sites, i.e. the positions in space, from which records are obtained. Sites are usually defined by reference to spatial co-ordinates but may relate directly to the position of an organism (or part thereof) which may or may not be actually included in the sample, or even the population, under study, e.g. insects per leaf, seedlings in relation to parent trees.

The strategy has a profound effect on the choice and limitations of subsequent data processing procedures. Decisions at this stage include, for instance, whether the samples should be distributed at random, systematically or a combination of both; whether the samples should be taken from one, two or three dimensions; or whether they should be plotless or consist of quadrats.

2 *Sampling Techniques* This stage covers the mechanics of obtaining

plant or animal records from the sample sites. These records may be merely records of 'presence' or some measure of quantity obtained either as absolute values or some estimate derived from sub-samples at each site. Techniques are generally less restrictive than strategies on subsequent data processing. Nevertheless, the type of measure used and the degree of sub-sampling will affect the statistical reliability of the results.

In plant ecology, decisions at this stage are related to the choice of vegetation feature to be recorded. For quantitative records, percentage cover, shoot density, and measures of bulk are commonly used.

In animal ecology the choice of spatial and temporal characteristics to be recorded, the use of baits in traps, marking of individual animals, length of time between samples, duration of a sweep sample etc., is also important.

In this section the exercises aim to provide students with situations which will focus their attention on some of these problems. The selection of which population feature to record and the size of sample required for reliability of the data are examined in some detail.

Exercises 1 and 2

The number of sampling units

Principles

If a large number n (> 30) of sampling units is taken and the numbers of a plant or animal counted in each, it is possible to erect confidence limits around the mean number/sample. Although the true population mean (μ) is constant, the value of the sample mean ($\bar{x}$) varies from one sample to the next. However, whatever the actual nature of the distribution of x (the individual counts), the *central-limit theorem* tells us that $\bar{x}$ is distributed approximately normally around μ.

The standard deviation of the sample means (or standard error as it is often called in this context) can be estimated as

$$\text{SE of sample mean} = \sqrt{\frac{\text{variance of } x}{\text{number of sample units}}}$$

$$\text{i.e.} \quad SE_{\bar{x}} = \sqrt{\frac{\sigma^2}{n}}$$

When the sample size is large (> 30), the distribution of sample means approximates to normality with a mean μ and the standard error can be estimated from

$$SE_{\bar{x}} = \sqrt{\frac{s^2}{n}} \quad \text{where } s^2 \text{ is the variance of the sample.}$$

Then from tables of the normal distribution the 95% confidence limits are given by

$$\bar{x} \pm 1.96\ SE_{\bar{x}}$$

A full account of the estimation of sample accuracy can be found in most statistical texts. Here the emphasis will be on the estimation of the number of sample units required to reflect accurately the population mean. Two examples investigating this aspect are considered: the calculation of the necessary sample size from a pilot sample of insects and the calculation of sample size in a plant community to ensure a representative picture of its species composition.

In the laboratory exercise, for a pilot sample of an insect population the number of samples required (N) may be obtained from the following formula

$$N = \frac{4s^2}{D^2\bar{x}^2}$$

where s^2 = variance of the pilot counts, $\bar{x}$ = the mean and D = the relative error in terms of percentage confidence limits of the mean. For example, if we are satisfied with 95% confidence limits with a range of ±40% of the mean (i.e. a standard error of *c*.20% of the mean), then $D = 0.4$. The sample size necessary to give an estimate of the population mean within ±40% of the true value is given by the above formula with $D = 0.4$. The formula then simplifies

$$N = \frac{4s^2}{D^2\bar{x}^2} = \frac{4s^2}{0{\cdot}4^2\bar{x}^2} = \frac{25s^2}{\bar{x}^2}$$

See Elliott (1977) for further explanation.

This formula enables the required number of samples to be estimated for a wide range of ecological situations both in animal and plant studies, providing it can be assumed that the data are derived from a homogeneous environment. Although N will vary markedly through the season, it is not usual to calculate it on each sampling occasion so a rough initial estimate is all that is required.

Calculating the number of sampling units required in plant ecology presents similar problems. Where quadrat methods are used to record density or performance it is usual to construct a travelling mean from a pilot survey of 100 quadrats. The mean is calculated for the first 10 quadrats, 20 quadrats, 30, 40 etc. and plotted against quadrat number. Graphs constructed this way resemble that shown in Fig. 1/2.1. From the shape of the graph it is possible to determine when the mean is a reliable value; it is usual to stop sampling when the graph fluctuates within 10% of a mean value.

When, on the other hand, the requirement is to establish whether the sample has accurately reflected the species composition of a community a different approach is necessary. One of the possible approaches is based on species-area curves which are used to define minimal area—the smallest area which can contain the characteristic species complement and structure of a plant community. The concept of minimal area is essential to the understanding of vegetational homogeneity and an important concept for the classification and mapping of vegetation (Greig-Smith, 1964; Hopkins, 1957; Shimwell, 1971). A more recent approach uses the concept of the minimal quadrat number which is obviously related to area. Empirical investigations have demonstrated that the relationship between species and quadrat number provides curves similar to species-area curves and further that the equivalence point (i.e. when the number of species and number of quadrats equate) on the curve can be used to define the minimum number of quadrats required. The curve below the equivalence point represents the zone in

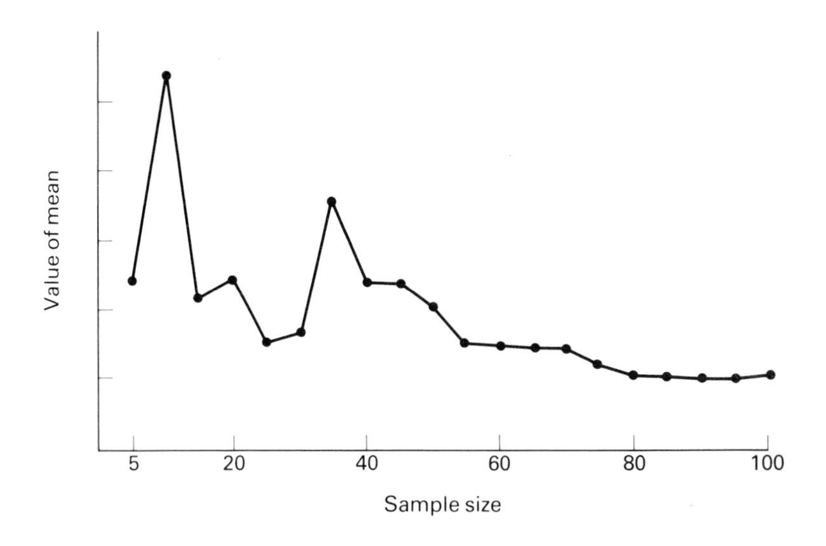

Fig. 1/2.1 A travelling mean demonstrating the reduction in the variation as the sample size is increased (after Kershaw, K. A. (1973). *Quantitative and Dynamic Plant Ecology*, 2nd edition. Edward Arnold, London.).

which species number increases rapidly with respect to quadrat number; above the equivalence point the quadrat number increases rapidly with respect to species number.

Exercise 1
Sampling a population of unknown density

Apparatus

For each student pair: jam jar with screw top containing *Tribolium* (or any small beetle) and *c.* 500 g of finely-sieved (0.5 mm mesh) wholemeal flour; sampling spoon—this can be constructed from a plastic medicine spoon with the bowl bent at right angles to the stem after softening by immersion in boiling water; 0.5 mm and 0.75 mm diameter mesh sieves; paint brush size 00; 2 Petri dishes; single-pan balance (accuracy 0.01 g); large beaker (*c.* 750 ml); plastic cup.

Preparation

From a large stock culture of *Tribolium* transfer several hundred beetles to each jam jar and add flour to half fill. A range of densities from approximately 250 to 1000 beetles per jar has been found suitable.

Procedure

1 Gently roll and invert jar to homogenize the cultures.
2 Weigh the beaker or use the tare on the balance to bring the reading to zero.
3 Carefully transfer culture from jam jar to beaker and reweigh the beaker.
4 Remove singly 20 samples with the sampling spoon—level spoons give an acceptably constant sample—to a Petri-dish where the number of beetles (x) is recorded. Flour and beetles are returned to the culture and mixed after the last sample.
5 Calculate required sample number for 40% (0.4) level of accuracy from

$$N = \frac{25s^2}{\bar{x}^2}$$

$$\bar{x} = \frac{\Sigma x}{n}$$

$$s^2 = \frac{1}{n-1}\left(\Sigma x^2 - \frac{(\Sigma x)^2}{n}\right)$$

where x = no. of beetles taken/sample and n = no. of samples.
6 Using the value N continue sampling until the total number of samples reaches N, if there is enough flour to do this.
7 Calculate the new $\bar{x}$ value.
8 Weigh 20 samples and calculate mean weight per sample.
9 Calculate the total number of samples contained in the culture from

Total Samples (T) = Weight of culture/mean weight of sample.

10 Population size is estimated from $T\bar{x}$.
11 Sieve the whole culture and count the actual population P.
12 Compare the true population size with the estimated size ($T\bar{x}$) and the sample mean (P/T) with the estimated sample mean ($\bar{x}$) (from 7). Is the estimated sample mean within 40% of the sample mean? Would a few large samples be more accurate than several small ones?

Exercise 2
The effects of quadrat number on the sampling of vegetation

Apparatus

For each student pair: m^2 quadrats; metre rules.

Preparation

If a large class (> 20 students) is used for the exercise then a wide range of vegetation types may be investigated but to investigate some of the problems of quadrat sampling five types of vegetation will provide sufficient information. Vegetation types found suitable are grassland (including ornamental lawns), heathland, woodland, bog, salt-marsh, sand dunes and marginal vegetation, i.e. woodland verges, lakeside vegetation etc. It should be possible to arrange for the field exercise to site groups of students in 5 different vegetation types without excessive travelling.

Procedure

1 Students work in pairs. For the given vegetation type, using the m^2 quadrats, identify and record all the species in each of 100 randomly-placed quadrats. Where identification is difficult the species may be coded and a sample taken for later identification.

2 At the same time as recording the number of species, a single common species is selected—preferably present in all quadrats. Students decide on a particular morphological attribute of the species, e.g. height, number of flowers, individuals per quadrat. The attribute selected will depend on the species but should always be easy to measure; it should be recorded for each quadrat.

3 On return to the laboratory, plot the species–quadrat number curve from 1 above and determine the equivalence point. Plot the running mean of the selected measure in 2 above for 10, 20, 30, 40, 50, 60, 70, 80, 90, 100 quadrats. In addition, calculate and plot the standard error of the mean (expressed as a percentage of the mean) against quadrat number (Greig-Smith, 1964, p. 33).

4 Collate the class results and compare the results within a community and between communities. Take note of any differences between the species–quadrat number curves for the different communities. What do these differences tell us? What factors could influence these curves? Are there any drawbacks at stopping at the equivalence point?

The measurement of a single attribute of the vegetation raises the questions of: When has a sufficient level of accuracy been attained? Why do some attributes require more samples than others to estimate their value to a given level of accuracy? What would be the effect of smaller and larger sampling units?

If these and other relevant questions are considered, the selection of appropriate sampling procedures in student project work can be facilitated.

Discussion and conclusions

In these exercises the interpretation and significance of the findings will require some background reading in both books on quantitative ecology and more general statistical works. Both the laboratory and the field investigations are concerned with the size of sample (i.e. the number of sample units, as used by most text books). In many ways both exercises are examining the relationship between the sample mean ($\bar{x}$) and the true population mean (μ). Laboratory exercises have an advantage in that an absolute value for μ can be preset or readily determined. In contrast field exercises enable a wide range of variables to be examined, often at the discretion of the students themselves. Results from the field exercise are useful as a concrete base for additional work on experimental design.

Consideration of the species–quadrat relationships in the field exercise should involve the concepts of homogeneity and pattern (see Section 2) as well as those of species richness and diversity (see Section 5). Criteria other than the equivalence point might be considered for the selection of a suitable number of quadrats using the species–quadrat number curve.

Sampling efficiency should be discussed with regard to levels of accuracy required for the study objective, redundancy in data and the precision of the method of data collection, e.g. the physical method used to measure leaf length is only accurate to a percentage of the mean length per sample and the measured weight of biomass samples can only be as accurate as the balance used. For instance it is useless trying to obtain very high accuracy for the sample mean when the methods used to measure the attribute may not be able to attain the same level of accuracy.

Further investigations

1 A non-homogeneous environment can be substituted for the *Tribolium* culture in the laboratory exercise by allowing the beetles to stratify in the jar overnight. Comparisons between results of sampling homogeneous and non-homogeneous cultures allow the underlying assumptions of some statistical techniques to be investigated.

2 Confidence limits can be erected regardless of the distribution of the counts per sample, which may follow, for example, a Poisson or a binomial frequency series, since the sample means will form a distribution approximating to normality as the number of sample units increases (p. 5). The relevance of this feature can be explored by sampling highly patterned phenomena such as shingle vegetation.

3 Sampling in freshwater habitats causes many problems. A wide range of topics suitable for projects can be based on the estimates of population size of, for instance, small crustacea, e.g. *Daphnia* or *Cyclops*, or insect larvae such as mayfly and dragonfly larvae. Many useful techniques of direct application to freshwater habitats may be found in Elliott (1977) who also describes the mechanics of collecting the data. Calculating total numbers of organisms in freshwater habitats is often impossible and density measures, e.g. numbers/m^3, are used instead. However, care should be taken to account for vertical stratification—a phenomenon often overlooked by students.

Cultures of freshwater organisms also provide a good source of laboratory practicals with the advantage that absolute population values can be derived.

4 When a plant community is sampled by quadrats and the density of every species is required, an interesting concept relating to sampling efficiency arises: the common species require fewer sample units for an accurate (at any defined level, e.g. 10%) estimation than do rare species (Greig-Smith, 1964). Sampling density in a community could theoretically employ a system when recording of the common species is stopped after a calculated number of quadrats and only the rarer species recorded in subsequent quadrats. Using methods of predicting the necessary sample number from formulae (see p. 6 and Greig-Smith, 1964) then both species-rich and species-poor areas could be investigated with optimal efficiency.

Exercises 3 and 4

The effects of quadrat size

Principles

The choice of quadrat size is critical to any sampling scheme. Whether the quadrats used are of some geometrical shape or are point quadrats, the actual dimensions will affect the accuracy of estimates and sometimes the further analysis of these results. In Exercise 2 the effect of quadrat size on the estimation of density was investigated, and in the exercise on sample size (i.e. number of sample units) the relationship between increasing quadrat size and standard error was examined. This section concentrates on two aspects of quadrat size: the effect of quadrat size on the analysis of plant associations and the effect of the size of a point quadrat when used to estimate vegetation cover.

The first of these exercises demonstrates the reason why plant associations must be carefully interpreted, for although two species may be positively associated at one level (i.e. scale) they may be negatively associated at another. This feature is obvious if one considers the fact that no two plants can occupy the same co-ordinates in space yet may be obligate associates. Similarly, two calcicole species may be negatively associated in that they never are found close to each other, for instance within the same $1m^2$ quadrat, perhaps due to allelopathy or some other competitive interaction (see Section 4), yet when considered on the scale of chalk grasslands they may be regarded as constituent parts of one community.

To demonstrate that this effect can operate within a community the laboratory exercise uses paper models of plant associations to investigate the effect of quadrat size on the resulting analysis of the associations.

Point quadrat recording in the form of frames of pins is a much used method for collecting vegetation data. It is used to estimate cover, bulk and species composition. The major drawback in using these point quadrats is the length of time it takes to collect the records. However, if they are collected carefully they can objectively represent the above vegetation attributes with a high degree of accuracy. Theoretically the point quadrats are a representative sample of the infinitely large number of possible points within a given area and it is because of this fact that the closer the point approximates to infinitely small the more accurate will be the estimates. Similarly, the larger the number of point samples the better they represent the true vegetation structure.

In the field exercise different pin diameters are used in pin frames to investigate how this alters cover estimates for a range of vegetation types.

Exercise 3
The effects of quadrat size on plant associations

Apparatus

For each student: a 25 × 25 cm simulated community (see Preparation below); wire quadrats of 1 × 1, 2 × 2, 3 × 3 cm or alternatively three sheets of acetate 25 × 25 cm marked with 50 random quadrats of the appropriate size.

Preparation

There are many ways in which simulated plant communities may be derived and a wide range of suitable materials for their construction exists. A simple situation can be modelled by a large sheet of graph paper onto which are stuck self adhesive circular coloured labels *c.* 1 cm in diameter in set positions. These positions are determined by the organizer to provide a different situation for each pair of students. Typical arrangements are: 'random', i.e. positions of paper 'species' A and B are alternately determined by random co-ordinates; 'random, strongly positively associated' where at each random co-ordinate species A is selected and species B placed so that their peripheries touch; 'clumped, strongly positively associated' where each random co-ordinate is used to define the centre of a circle in which equal numbers of A and B are placed at random, e.g. a circle of 8 units diameter containing 4 × A and 4 × B of 1 unit diameter placed at random so that they may touch but not overlap; 'associated' where A and B are placed within set distances of each other, e.g. 2 cm; 'negatively associated' where A and B are located with a preset minimum for the distance between them, e.g. 5 cm; and 'clumped, negatively associated' where A and B are located in clumps of only A or B. Many more combinations can be devised and each one at different densities.

For the exercise it is best to use a range of set densities, e.g. three densities: high (50% cover), medium (25% cover) and low (10% cover), for each pattern. Mark the derivation of the arrangement on the reverse of each model community.

Procedure

1 Each student takes a particular community pattern at one of the three densities.

2 Using the wire quadrat (or acetate overlay) 50 randomly-placed

quadrats are sampled and the data for the selected pattern is recorded as below

No. of quadrats containing only A	c
No. of quadrats containing only B	b
No. of quadrats containing both A & B	a
No. of quadrats without A or B	d

3 Calculate the association between species A and B from a 2×2 contingency table as follows. Let the total number of quadrats used (50 in this case) be n, then from 2 above

'Species' B	'Species' A +	'Species' A −	
+	a	b	$(a+b)$
−	c	d	$(c+d)$
	$(a+c)$	$(b+d)$	n

4 A χ^2 test is now applied to the results of the 2×2 contingency table to test for association. Further details can be found in Kershaw (1973). This formula includes Yates' correction which overcomes the problem of an expected number less than 5.

$$\chi^2 = \frac{(|ad - bc| - \frac{1}{2}n)^2}{(a+b)(c+d)(a+c)(b+d)}$$

The vertical lines mean that we have to take the absolute (i.e. positive) value of the difference between ad and bc.

5 Refer to χ^2 tables (Parker, 1979) for 1 degree of freedom and determine whether the value obtained is significant at the 5% level. If it is, this tells us that the difference between the observed and expected values is unlikely to have occurred by chance fluctuation.

6 To establish the direction of the trend (if any) of association found in 5 we use the formula

$$\left(\begin{array}{l}\text{Observed number of}\\ \text{joint occurrences}\end{array}\right) - \left(\begin{array}{l}\text{Expected number of joint occurrences}\\ \text{assuming no association}\end{array}\right)$$

$$a - \{(a+b)(a+c)/n\}$$

If the result is positive then the species A and B are positively associated and if negative then they are negatively associated *for the quadrat size used.*

7 Repeat 2–6 for the other quadrat sizes so that each student has completed the analysis using all three quadrat sizes on a selected 'community'.

8 Tabulate the class results for each pattern type.

Association between species A and B

Density	Quadrat size		
	1 × 1 cm	2 × 2 cm	3 × 3 cm
Low			
Medium			
High			

Use + = positive association
– = negative association
* = significant at 5% level

Exercise 4
The effect of the diameter of point quadrats on cover estimation in herbaceous vegetation

Apparatus

A pin frame for each student pair. For many field situations a useful size is one that holds 10 pins at 5 cm intervals. If the distance between the last pin and the centre of the supporting leg (see Fig. 3/4.1) is 2.5 cm then the whole frame can be pivoted on one leg to provide contiguous readings. Likewise, if the frame is 5 cm wide, frames can be placed adjacent to each other to cover a 10 × 10 pin grid. The frames can be constructed from wood or hollow aluminium with brass sleeves to hold the pins and a built-in spirit level to maintain a horizontal plane. The pins may be constructed from bicycle spokes, stainless steel rods or welding rods. The harder the material used the better since soft materials such as copper bend too easily. It is better to use a thick rod (e.g. the bicycle spoke) to which a finer rod or sharp point is added rather than a thin rod which will flex in the wind and be too whippy to be reliable. For this experiment, pins ranging from a needle point (infinitely small) to a pin with a 5 mm diameter tip, are required to provide at least 5 different point quadrat sizes, i.e. infinitely small, 0.5 mm, 1 mm, 2.5 mm, 5 mm.

Preparation

1 The areas found useful for this experiment have a high percentage cover, i.e. 80–100%, and contain three or four major species. Grassland, heathland, saltmarsh, sand dunes and woodland ground vegetation have

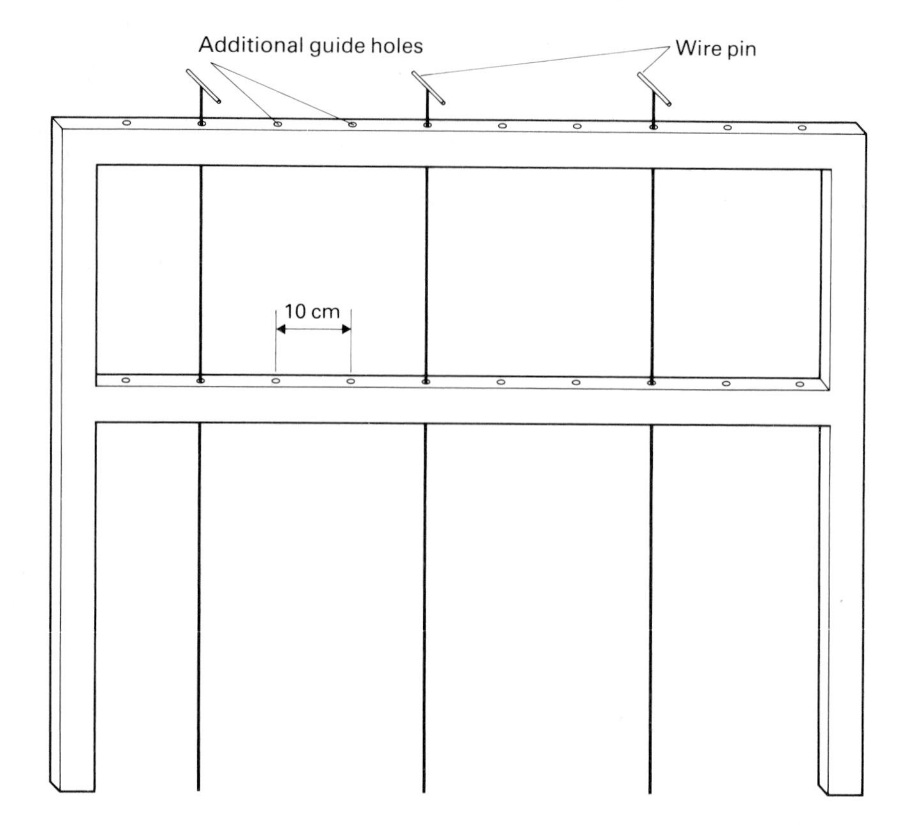

Fig. 3/4.1 A pin frame or point quadrat (after Mueller-Dombois, D. and Ellenberg, H. (1974). *Aims and Methods of Vegetation Ecology.* Wiley, New York.).

all been used for this experiment. It does not matter how students arrange the pin frames within a chosen community but for comparative studies it is obviously sensible to be consistent, i.e. all students arrange the pin frames in a transect or in a grid.

2 Before the experiment or during the travelling time familiarize the students with the plants they will find.

Procedure

1 Each pair of students marks an experimental area in the chosen plant community; either a transect of 10 pin frame lengths (5 m) or a grid of 10 parallel frames.

2 Each pair of students is supplied with a frame with a different size of pin, from the range given in the Apparatus section, and records the first

species touched by the pin as it is moved from the top of the plant canopy to the ground, i.e. 1 record per pin. Where no species are touched, bare ground is recorded.

N.B. This arrangement of students demands a class of 10+ students to operate. An alternative for smaller classes is for each student to work on a single pin frame location but to use all 5 sizes of pin.

3 Collect the class results for each species as below

Species:

Pin diameter (mm)	% cover
Infinitely small = x_1	y_1
0·5 = x_2	y_2
1·0 = x_3	y_3
2·5 = x_4	y_4
5·0 = x_5	y_5

Percentage cover estimates for each pin are determined from

$$y_1 = \frac{\text{no. of pin hits for species A}}{\text{no. of pins of } x_1 \text{ dia. used}} \times \frac{100}{1}$$

4 Plot the graph of % cover against pin diameter and estimate the regression from

$$y = a + bx$$

where $$b = \left(\Sigma xy - \frac{(\Sigma x)(\Sigma y)}{n}\right) \Bigg/ \Sigma x^2 - \frac{(\Sigma x)^2}{n}$$

and $$a = \bar{y} - b\bar{x}$$

In this experiment the b values tell us the slope of line, which is steeper the larger the b value. Therefore the higher the b value the greater the effect of pin diameter (see Fig. 3/4.2).

Discussion and conclusions

Many field experiments can, on analysis, produce spurious results which are difficult or impossible to interpret ecologically. The laboratory exercise demonstrates one way in which inconsistencies in results may arise. Students often overlook the effects of quadrat size on their results and in this exercise the difficulties facing the analysis of plant associations are revealed. Discussion of the results of these experiments will usually lead students to an awareness of the difficulties and a commonsense attitude towards future sampling decisions. A most

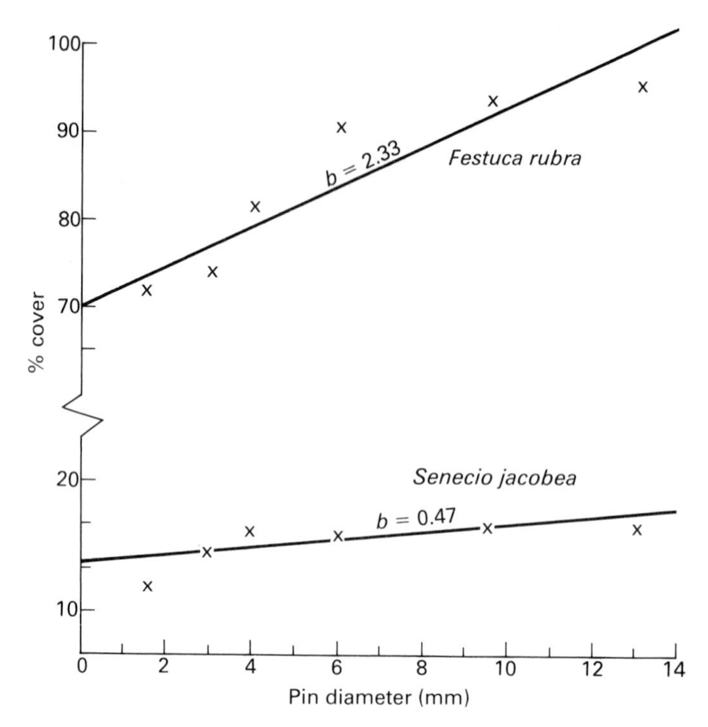

Fig. 3/4.2 A comparison of the exaggeration of cover values with increasing pin diameter for two species.

thorough consideration of the problems associated with quadrat sampling is found in Greig-Smith, 1964.

Aspects of quadrat sampling (point quadrats) are also explored in the field exercise. Here the relationship between pin diameter and cover estimation is investigated. The results invariably show an increase in cover estimation for plants as pin diameter increases, with a corresponding decrease in the value for bare ground. Since the theory of point quadrat (pin frame) sampling is based on the acquisition of a representative sample of the infinite number of infinitely small points in any area, it follows that the larger the deviation from infinity the larger the expected error, for a given number of point quadrats.

Once the relationship between pin diameter and cover estimate has been established, in general terms, it is possible to extend the study to explore the reasons for differences in the regression coefficient for different species. Here it is possible to examine the life forms of the species in relation to their regression coefficients. Likewise leaf area indices and broad categories of plants, e.g. monocotyledons/dicotyledons, can be compared using the *b* values.

Further investigations

1 As an alternative to square sampling units (quadrats) it is also possible to examine the efficiency of differently shaped units, e.g. rectangular, circular, hexagonal, triangular. These may be used in exercises in the laboratory with paper or photographic data and for field investigations.

2 There is no theoretical reason why point quadrat sampling for cover estimation should be confined to the pin frame method. Randomly arranged pins can be used to estimate cover and there is some evidence to support the use of this method (Greig-Smith, 1964) since fewer pins are required for any given level of accuracy. The difficulty in deciding between the two methods is complicated by the fact that in some vegetation it takes longer to arrange random pins than pins in a frame. This feature may, in terms of overall efficiency, make the pin frame more desirable.

A number of interesting projects could be based on the comparison of the two methods of point quadrat sampling taking account of, for instance, man hours required to collect data, accuracy of data and the influence of different vegetation types on these results. One often requires an overall cover estimate for a large area of vegetation, e.g. *Spartina* sp. on a salt marsh; here we need to know whether this would be best estimated by x random pins or $x/10$ pin frames of ten pins each and how the pin frames themselves should be located.

3 Point quadrats are used mainly in plant ecology yet there are situations in animal ecology which are amenable to this type of sampling. The obvious limiting factor is the mobility of animals and it is only in situations where this movement is minimized (i.e. sedentary or sessile animals) that this type of sampling can be used. The seashore is the habitat offering the widest possibilities for point quadrat sampling and the method is frequently used to estimate barnacle cover and the distribution of different barnacle species (see also Exercise 34). Similar exercises to those described above could be applied to both the animals (particularly molluscs) and the algae found in this habitat.

Exercises 5 and 6

Quantitative measures of vegetation

Principles

Two of the most valuable vegetation measures are cover and density. Cover is the proportion of ground, usually expressed as a percentage, covered by the horizontal projection of the aerial parts of a plant. Density is defined as the number of plants (or plant units such as shoots) per unit area. Although these measures are often difficult to estimate they are usually less time consuming than the collection of biomass or yield data, less damage is done, and they are more easily interpreted than measures of frequency. The latter are simple to collect but difficult to interpret biologically since a multitude of both sampling and vegetation characteristics may alter their value.

Cover is a concept that requires close consideration for it is often used to mean top-cover in which case only the uppermost layer of horizontal vegetation is considered and the values for all species and for bare ground added together will equal 100%. Alternatively cover may be used to describe all horizontal extensions of the plants present (remembering that even a vertical blade of grass has a finite horizontal area); in this case the summed value will be in excess of 100% in all but sparsely vegetated areas. It is the former concept we are interested in here but the existence and ecological significance of the latter, particularly in areas with dense cover and much overlapping and layering of the vegetation, should not be overlooked.

In the field exercise various methods of measuring and estimating vegetation cover will be explored.

Density is usually more straightforward to understand but as a measure usually requires qualification. For instance, density measures may concern whole plants in the case of forest trees or tillers in the case of grasses. For many species shoot density is a useful measure, particularly if the study requires estimates of performance in relation to some habitat factor. In all cases the selection of density measures must be related to both the morphology of the species under investigation and the study objectives.

The reciprocal of density is related to the mean area of the plants and thus density is related to the mean distance between individuals. In the laboratory exercise the calculation of density is based on both this relationship and direct measurement by quadrat sampling.

Exercise 5
Quantitative estimates of density

Apparatus

Per student pair: 1 m^2 pin board covered with 1 mm^2 graph paper; panel or mapping pins *c.* 30 × 1 mm; a 5 × 5 cm (25 cm^2) and a 7.1 × 7.1 cm (50 cm^2) wire quadrat; wristwatch or clock; paper labels for pins.

Preparation

The pins are given sequential numbers on paper labels and pushed into the board at random co-ordinates (from tables) so that each student pair uses a board to produce a random arrangement of pins at a density within the range of 50 to 500/m^2. Label each board with its pin density.

Procedure

Students should time themselves for the data collection stage of each method of density estimation.

1 Using the quadrats provided:

(*a*) for the 5 × 5 cm quadrat 60 random co-ordinates are read from random number tables (00–99) to locate the position of the bottom left hand corner of the quadrat on the pinboard. The number of pins enclosed each time is counted and the mean number of pins per quadrat is calculated (n_1).

(*b*) for the 7.1 × 7.1 cm quadrat 30 random co-ordinates are read from random number tables to locate its position as above. The mean number of pins per quadrat is calculated (n_2).

2 Data for the quadrat methods are recorded as below

	Quadrat size	
	25 cm^2	50 cm^2
Mean number per quadrat	n_1	n_2
Variance		
Standard error of mean		
Estimate of the no./m^2		
Standard error of estimate		

For these calculations, see Exercise 1 or Parker (1979).

3 Locate 50 random pins (using random number tables to identify each one) and measure the distance from each one to its nearest neighbour in cm. Duplicated random numbers should be ignored. Calculate the mean distance $\bar{d}_1$ between nearest neighbours, and the

density per cm^2 from

$$D_1 = \frac{1}{(2\bar{d}_1)^2}$$

4 Locate 50 random co-ordinates from random tables and measure the distance to each one's nearest pin in cm (to save time use the co-ordinates previously used) (see Fig. 5/6.1).

5 Calculate the mean distance in cm ($\bar{d}_2$) between a random co-ordinate and its nearest pin, and the density from

$$D_2 = \frac{1}{(2\bar{d}_2)^2}$$

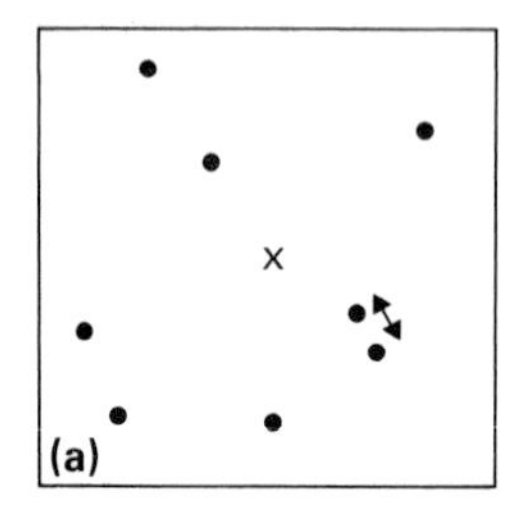

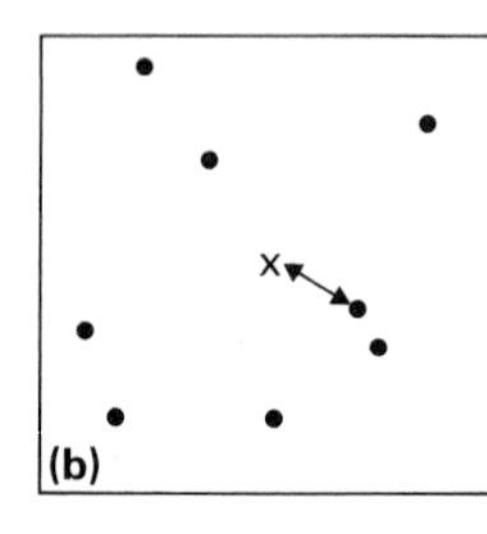

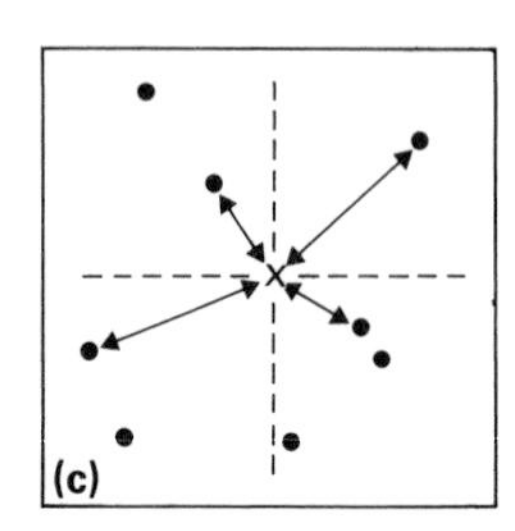

Fig. 5/6.1 Plotless sampling methods: (a) nearest neighbour method; (b) closest individual method; (c) point-centred quarter method.

6 Use 25 random co-ordinates (again the same set as in 3 can be used) to locate the centre of a cross which is aligned parallel to the edges of the board. Record the distance (cm) to the nearest pin along each of the four lines of the cross. Calculate the mean density (no./cm^2) of *all* such distances ($\bar{d}_3$), and the density from

$$D_3 = \frac{1}{(\bar{d}_3)^2}$$

7 Express the results of the nearest-neighbour methods as density per m^2 (i.e. multiply by 100^2).

8 Tabulate all density measures and compare the accuracy of the estimates and time required for data collection.

Exercise 6
Cover estimation in vegetation

Apparatus

$1 m^2$ quadrat; $1 m^2$ quadrat with grids 10×10 cm; $1 m \times 10$ cm pin-frame; wooden pegs (e.g. $2 \times 2 \times 20$ cm); wristwatch.

Preparation

This exercise has been found suitable for many types of vegetation including shingle, grassland (lawns, playing fields, etc.), heathland, and woodland ground flora. To stimulate a wider discussion of the results the selection of more than one community to provide a range of species and life forms is recommended. Where this is possible groups of students are allocated to each community and follow the same procedure as below, but in addition they compare results between communities and discuss the results in terms of differences in vegetational composition between the communities.

Procedure

Students time themselves for the data collection phase of each method.

1 Each pair of students marks a $1 m^2$ plot with pegs. Within the meter square top cover records are collected by the following methods:

(*a*) *Direct Estimation of top cover* Visual estimate of top cover for whole quadrat. Record each species to nearest percent. The total for all species and bare ground will equal 100%.

(*b*) *Sub-quadrat estimation of top cover* In 25 of the 100 $10 cm^2$ sub-quadrats (i.e. every fourth sub-quadrat) estimate the percentage cover of each species. Sum the results and calculate the mean to obtain an estimate of cover percentages for the $1 m^2$ quadrat. The sum of the mean species values including bare ground will equal *c.* 100%.

(*c*) *50% method* In the 100 sub-quadrats record the number of quadrats in which each species occupies $\geqslant$ 50% of the area. The summed values for this method often lie below 100% since many sub-quadrats will contain a species mix where no single species (or bare ground) will reach 50% cover.

(*d*) *Braun-Blanquet 5 point scale* Estimate visually the cover of each species and bare ground for the $1 m^2$ plot using the following scale

+	very rare	less than 1%
1	rare	1–5%
2	occasional	6–25%
3	frequent	26–50%
4	common	51–75%
5	abundant	76–100%

(*e*) *Domin scale* Estimate visually the cover of each species and bare ground for the $1 m^2$ plot using the following scale

+	A single individual
1	Scarce, 1–2 individuals
2	Very scattered, cover small < 1%
3	Scattered, cover small 1–4%
4	Abundant, cover 5–10%

5	Abundant, cover 11–25%
6	cover 26–33%
7	cover 34–50%
8	cover 51–75%
9	cover > 75%, but not complete
10	Cover practically complete, 100%

(*f*) *Pin-frame estimation of top cover* Using the pin frame parallel to and 5 cm from the top edge of the quadrat, then parallel to the edge at 10cm intervals, the holes will lie over the centres of 100 sub-quadrats (Fig. 5/6.2). Lower the pins singly and record the species first hit by each pin. Sum the results to give the percentage (100 pins total) cover of each species and bare ground; these figures will add up to 100%.

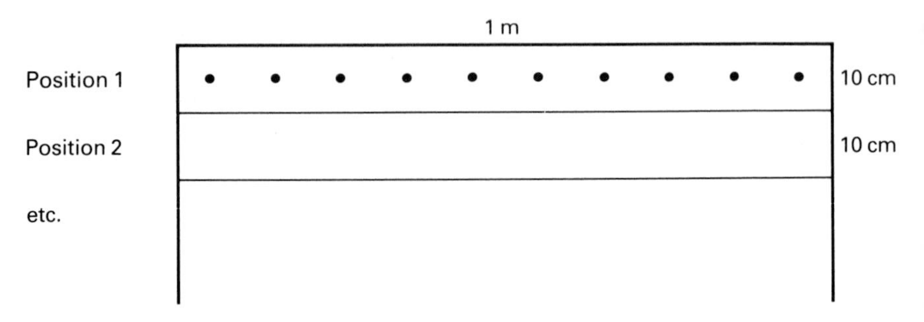

Fig. 5/6.2 Arrangement of pin positions in Exercise 6.

2 In all the methods (*a*)–(*e*) each estimation has a strong subjective bias. To reduce individual differences in estimation the values given to each species for these methods should be based on agreement between both members of a student pair. Since the operation of scaled estimates depends on practice for consistency, all the students should be either naive to their operation or equally practised, e.g. allowed 5 pilot estimates on random sites before the test quadrat.

Pin-frame estimates are the most objective of those used in this exercise but students should take care to ensure that records are taken of the species touched by the points and not the sides of the pins.

Discussion and conclusions

Density estimates are often required in vegetation studies and the selection of the appropriate sampling technique can influence both the efficiency of the study and the reliability of the results. In the laboratory exercise a number of important points are raised which can be discussed in relation to the class results. Firstly, there are questions related to the direct comparisons of the methods. Which method is most accurate? Are some methods easier to operate than others? What restriction would different field conditions (forest trees, grasses, woody shrubs etc.) impose on their application? Are some more time-consuming than others? If so, would spending more time on the quicker methods (i.e. more quadrats or density measures) improve the accuracy for a given effort?

Secondly, there are questions relating to the methodology of the individual sampling strategies. Does the size of quadrat affect accuracy? How do 'edge effects' (i.e. measurement at the borders of the study area) influence the methods? Is it possible to locate accurately quadrats on the selected co-ordinates or does the shape and form of individuals impose restrictions on quadrat position? Do the estimates of the density alter in accuracy depending on the absolute density value?

In populations provided for the laboratory practical the individuals were arranged at random, which is rarely the case in nature (see Section 2), and the accuracy of the results obtained from extremely patterned populations is likely to be severely affected. This is particularly true of the nearest neighbour methods which use formulae based on random arrangements. The formulae and their origins are discussed by Greig-Smith (1964, Ch. 2).

The estimation of cover is comparatively easy compared with the estimation of frequency, or density. Although most of the methods employed by ecologists, and represented in Exercise 6, are subjective assessments they work for most purposes. Where different sub-communities are compared so that different sizes and shapes of plants and various degrees of species richness are encountered, the benefits of one method over another are revealed. Since the pin frame technique merely takes a sample of the infinite number of possible points in an area, its accuracy should only relate to the number of points sampled and not the amount of cover of any one species. Subjective methods usually become less reliable as the cover decreases as most people overestimate small scattered units of cover. To overcome this problem two of the methods used, namely the Domin scale and the Braun-Blanquet 5 point scale have unequal steps scaled to compensate for this bias. Sub-quadrat methods attempt to improve the estimate of overall cover in each 1 m^2 quadrat by concentrating attention on smaller areas. Where a complete estimate is taken, smaller species and scattered cover are probably estimated best, but in the 50% method these will be missed altogether. However, if the primary interest is the dominant ground cover and speed is important then the 50% method becomes an attractive alternative.

Further investigations

1 One interesting aspect of collecting vegetation records is the variation between operators. One way of investigating this phenomenon is to arrange for a number of students to estimate the cover of the same 1m^2 area. This is not easy to manage and the physical disturbance by

one student may alter the cover for the next. Nevertheless these problems are overcome if vertical photographs are taken of the vegetation from *c.* 2m above the ground. A simple gantry is made to hold a 35mm single lens reflex camera which is loaded with fine grain black and white film. Numerous enlargements can be made of a selected picture to provide one per student. A white $1m^2$ quadrat frame is used to delineate the area on the ground and this can later act as a standard scale for photographic enlargement purposes. Heathland and shingle vegetation have been found suitable for this technique.

All the methods of cover estimation can be applied to the photograph and comparisons between them made.

2 An extension of the photographic cover estimation experiment would be to estimate the degree of improvement operators attain with increased practice. This work would involve a large number of short experiments using different operators and would be suitable for a student project. Variation between operators could then be examined in relation to their experience of a particular method of cover estimation.

3 In dense vegetation, species overlay each other and the total cover of plant material is usually in excess of 100% whichever method is used. Woodland vegetation, for instance, often has a stratified herb layer and affords a valuable opportunity to examine this feature. Here the plants need to be moved around during examination to reveal underlying species. In many woodlands, *Hedera helix* (ivy) is represented in nearly every $1m^2$ quadrat, yet during the spring and early summer it is completely obscured by taller species, leading to large underestimates of its true cover if only top cover is estimated.

All the cover methods can be used to examine this type of vegetation and some will be easier to operate whilst others will provide more reliable results. No absolute value exists with which results can be compared but pin frame methods, if a very fine pin is used (see Exercise 4), should provide the most reliable estimate. In the case of overlying layers of vegetation, pin frame estimates are collected by lowering the pins through the vegetation and recording all the species hit by the point at successive depths. This gives a cover value for each species.

4 Density estimates in the field are usually obtained from non-random situations. The effects of increasing aggregation on the accuracy of estimates may be examined in the laboratory by erecting model communities with random pattern centres. These are used to define the centres of aggregations, e.g. 2cm diameter, within which a set number of random points is arranged. By altering the proportion of the total population which is within these aggregations, as well as the size of the aggregated units and the population density it should be possible to determine the effects of aggregation patterns on the various methods of estimating density. Similarly, regularity could be simulated by placing a set proportion of the otherwise random population on the co-ordinates of an outline grid.

Exercises 7 and 8

Mark-recapture techniques for estimating the density of an animal population

Principles

Mark-recapture methods all depend on the assumption that when a sample of animals from a population is marked and allowed to redistribute among the unmarked ones the expected proportion of marked to unmarked animals in a subsequent sample is the same as the proportion of marked ones in the population. In the simplest mark-recapture methods (Exercise 7) several conditions have to be satisfied before the above assumption is justified. These are: (1) the marking method must not affect the animal; (2) the mark must last as long as the investigation; (3) the marked animals must become completely mixed in the population before the next sample; (4) the likelihood of an animal being captured must not change with its age; (5) the population is a closed one, i.e. no immigration or emigration occur, or if they do they can be determined; (6) there are no births or deaths during the investigation, or if there are, they can also be determined.

More complex methods do not require a static population (conditions 5 and 6) and can also allow for the death of animals during handling. Such a method is that developed by Jolly (1965). This method also differs from the others in that it is stochastic, that is it is developed on the basis that changes in numbers are a result of probabilities and therefore not precisely determined. A mortality rate of 50%, for instance, means that each individual has a 50/50 chance of dying rather than that exactly 50% of the population will die. Because of this, and because it does not assume that the probability of survival of an animal is constant, it is the most appropriate method for insect populations, where survival, birth and death rates change in a relatively short time.

The laboratory exercise uses the simple Lincoln Index, because in the short duration of the exercise we can make assumptions 1–6 fairly confidently. The field exercise uses Jolly's method because conditions 5 and 6 certainly do not apply to field populations. This method appears daunting but, although its theoretical basis is complicated, carrying out the exercise involves only simple, though fairly lengthy, arithmetic.

In the Lincoln (1930) method, the choice of criteria determining the point at which the second sampling should stop is important. Half the data should be collected as in (*i*) and the other half as in (*ii*) below.

(*i*) When the *total* number of insects counted reaches the number of originally-marked ones (already recorded), sampling should cease.

(*ii*) When the number of *marked* individuals captured reaches 20% of the total number of marked ones *known* to be present, sampling should cease.

The Lincoln analysis is based on the expected relationship

$$\frac{\text{Total population } (P)}{\text{Original number marked } (a)} = \frac{\text{Total in sample } (n)}{\text{Total with marks in sample } (r)}$$

1 If method (*i*) is used and if the value or r is *greater* than 20, the estimate of the total population is given by

$$\hat{P} = \frac{an}{r}$$

2 An estimate of the variance of the above estimate for P (var $\hat{P}$) is given by

$$\text{var } \hat{P} = \frac{a^2 n\,(n-r)}{r^3}$$

3 If method (*i*) is used but the value for r is *less* than 20

$$\hat{P} = \frac{a\,(n+1)}{r+1}$$

4 An approximate estimate for the variance (var $\hat{P}$) of the estimate is given by

$$\text{var } \hat{P} = \frac{a^2\,(n+1)\,(n-r)}{(r+1)^2\,(r+2)}$$

5 If method (*ii*) is used

$$\hat{P} = \frac{n(a+1)}{r} - 1$$

and

$$\text{var } \hat{P} = \frac{(a-r+1)\,(a+1)\,n\,(n-r)}{r^2\,(r+1)}$$

6 In the field exercise (Jolly), the data are recorded as below (the hypothetical example is necessary to make the method clear; in this case there were 7 separate samples, not one as in the laboratory exercise).

Date	Total captured (n_i)	Total marked and released (R_i)		Nos. recaptured from previous dates 1	2	3	4	5	6	m_i
1	96	95								
2	109	107		41						41
3	187	184		17	29					46
4	258	253		16	14	69				99
5	362	358		4	11	35	62			112
6	486	481		3	5	20	22	76		126
7	333	328		0	3	13	4	24	64	108
			r_i	81	62	137	88	100	64	

7 The basic formula for Jolly's method is

$$\hat{N}_i = \frac{\hat{M}_i n_i}{m_i}$$

where $\hat{N}_i$ = estimate of the population size on day i (i is any one of the sample days)
$\hat{M}_i$ = estimate of the total number of marked animals in the population on day i
n_i = total number of captures on day i
and m_i = total number of marked animals recaptured on day i.
A symbol with a cap (such as $\hat{N}_i$) means that it is an estimate rather than an absolute measure or count.

8 Like the Lincoln Index, it is argued that the expected proportion of marked animals in the sample is the same as the proportion in the population. We need, therefore, to estimate the *number* of marked animals in the population. On day i, this number is: those in the ith sample plus those *not* in the ith sample. The former is known, the latter is estimated from recaptures of marked animals on later dates.

9 In addition to the quantities in 7, we need:

R_i = number of animals *released* from the ith sample
r_i = number of the R_i animals which are subsequently recaptured
Z_i = number marked *before* day i which are not caught on day i but are caught subsequently
a_i = proportion of marked animals in the population on day i

Values for r_i, Z_i and m_i may be obtained from the above table by rearrangement, i.e. the total number of animals marked on day 1 which are recaptured $r_1 = 41+17+16+4+3+0 = 81$; $r_2 = 62$ and so on. To obtain Z_i, the figures in each *row* (horizontal), excluding the n_i and R_i columns, are added one at a time, and each new addition is recorded. When the data in the above table are added in this way, the first row is *17*, $+29 =$ *46*, the third is *16*, $+14 =$ *30*, $+69 =$ *99*, the fourth is *4*, $+11 =$ *15*, $+35 =$ *50*, $+62 =$ *112*, and so on. The rearranged table therefore looks like this

	1	2	3	4	5	6
	41					
	17	46				
	16	30	99			
	4	15	50	112		
	3	8	28	50	126	
	0	3	16	20	44	108
Z_{i+1}	40	56	94	70	44	
	Z_2	Z_3	Z_4	Z_5	Z_6	

The top figure in each of the new columns is the number of recaptures for the next day to the right (m_i) and is therefore a useful check on the values for m_i in the first table. The total for each column (*excluding* the value for m_i) gives Z_{i+1} so that the value for Z on day 4 (Z_4) is $50+28+16 = 94$.

10 We calculate the total population on day i as follows
Of the $(M_i - m_i)$ animals *not* caught on day i, Z_i are caught later. Similarly, of the R_i released on day i, r_i are caught later. As both these groups are assumed to have the same recapture rate

$$\frac{Z_i}{M_i - m_i} = \frac{r_i}{R_i}$$

By rearrangement, we can estimate M_i

$$Z_i R_i = (M_i - m_i) r_i$$

$$\frac{Z_i R_i}{r_i} = M_i - m_i$$

$$\hat{M}_i = m_i + \frac{Z_i R_i}{r_i}$$

As the total in the population must equal

$$\frac{\text{total marked in population}}{\text{proportion of population marked}}$$

i.e.

$$\hat{N}_i = \frac{\hat{M}_i}{\hat{a}_i}$$

and as the proportion of marked animals in the population on

$$\text{day } i = \frac{\text{total no. marked animals caught on day } i}{\text{total no. animals caught on day } i}$$

i.e.

$$\hat{a}_i = \frac{m_i}{n_i}$$

then

$$\hat{N}_i = m_i + \frac{Z_i R_i}{r_i} \div \frac{m_i}{n_i}$$

$$= m_i + \frac{Z_i R_i}{r_i} \times \frac{n_i}{m_i}$$

$$= \frac{m_i n_i}{m_i} + \frac{Z_i R_i n_i}{r_i m_i} = n_i + \frac{Z_i R_i n_i}{r_i m_i}$$

In the example above $\hat{N}_i$ for day 3 ($\hat{N}_3$) is

$$\hat{N}_3 = 187 + \frac{56 \times 184 \times 187}{137 \times 99} = 329$$

11 The standard error of this estimate is given by

$$\sqrt{\hat{N}_i(\hat{N}_i - n_i)\left[\frac{\hat{M}_i - m_i + R_i}{\hat{M}_i}\left(\frac{1}{r_i} - \frac{1}{R_i}\right) + \frac{1 - \hat{a}_i}{m_i}\right]}$$

which for the above example is 31.

The estimated number of grasshoppers in the population on day 3 is therefore 329 ± 31.

12 It is possible to calculate in addition

a) the survival rate from sample *i* to sample $i+1$;

b) the number of new animals joining the population between sample *i* and sample $i+1$;

c) the standard errors of these estimates.

The calculations are very lengthy (see Jolly, 1965 or Southwood, 1971) and are therefore not given here; nevertheless, they are worth calculating if a computer is available.

Exercise 7
Estimating the numbers of flour beetles (*Tribolium* spp.) using the Lincoln Index

Apparatus

For every pair of students provide: a wooden tray with wholemeal flour prepared as described in Exercise 13, but without a beetle culture; 50 glass specimen tubes (*c.* 8 × 2.5 cm) per student pair; cellulose or oil paint (any bright colour); 2 paint brushes (size 00 and 1); enough *Tribolium* adults to provide an average of at least 100/student pair; two Petri dishes each with the walls smeared with PTFE (I.C.I. Ltd.) (see Exercise 13), one with a moist filter-paper base, the other with dry paper; refrigerator; ice in a sandwich box; two empty sandwich boxes per student, the walls of each box smeared with PTFE; a 5 × 10 cm piece of card/student pair.

Preparation

1 Although the beetles can be marked by each pair of students, this would require a delay of at least an hour for the marked beetles to redistribute themselves among the unmarked ones. It is better, therefore, for the marking to be done one or two days before the exercise, the animals being added (when the paint is dry) to the flour with a *known* number of unmarked individuals as described below.

2 Release into each flour box a known number of beetles from the stock supply so that all the boxes receive different numbers all above 50; the higher the number the better.

3 To each box add a further 10% of beetles (but at least 20) which have been marked as follows.

4 Place the beetles a few at a time in the moist Petri dish, replace the lid and put the dish in the ice compartment of a refrigerator. Leave for *c.* 2 minutes or until the beetles have stopped moving; remove and place on the bed of ice.

5 Use the larger paint brush to turn the beetles the correct way up if necessary, then use the small brush to place a small dot of paint on one of the beetle's elytra (wing cases); avoid getting paint on any other part of the body.

6 When the paint is nearly dry, transfer the beetles to the dry dish using the larger brush and repeat 5 until the required number of beetles is marked (i.e. *c.* 10% of the number in the box).

7 Add the marked insects to the flour and write on the box the number of *marked* individuals added. Record elsewhere the total number in the box.

Procedure

1 Using random number tables to locate the sampling points in the flour, each pair of students should invert a tube and press it into the flour until it touches the bottom of the box. All the other tubes should be pressed into position in as rapid succession as possible to minimize disturbance of the beetles.

2 When all the tubes are in position, carefully remove them one at a time together with their contents; do this by sliding a piece of card beneath the tube mouth and lifting card and tube together.

3 Tip the tube contents into one of the sandwich boxes, count and record the number of beetles with marks and the total number. Leave them in the box.

4 Repeat 2 and 3, tipping the tube contents into the empty box each time and adding them to the second box following counting.

5 Calculate a, n, r and $\hat{P}$ according to the appropriate method in the Principles section. n in this case is the total number of beetles in all tubes combined and r is the total number with marks.

6 Each pair should enter their estimate for $\hat{P}$ and var $\hat{P}$ in a blackboard tabulation, 5 columns having been provided so that the following can be recorded
a) student pair *b*) method used (*i* or *ii*) *c*) estimate of $\hat{P}$ *d*) var $\hat{P}$
e) true value of P (entered by organizer).

Exercise 8
Estimating the numbers of insects in a field unit

Apparatus

Per pair of students: a small supply of artists' oil paint (a different colour for each marking occasion (see below)); a fine paint brush (size

00); a sandwich box with filter-paper in the bottom; a sweep-net or butterfly-net (the former is preferable; it is more robust than a butterfly net and is swept *through* the vegetation so that insects living there are knocked into it).

Preparation

1 Find a grassy area with a population of grasshoppers preferably dominated by one species. The area should have obvious boundaries (e.g. bounded by a road, hedge etc.). An ideal size is not more than half an acre (0.2 ha).

Procedure

1 Roughly divide the chosen area into sub-areas. Within each sub-area a pair of students should attempt to capture as many grasshoppers as possible (by direct observation as well as sweeping). This will be repeated for several successive days so the time spent collecting will be determined by the programme for the rest of the day; once the insects have been marked and released (see below) the exercise requires no further work on that day. Maintain flexibility however, as the weather affects the activity of the insects and more time will probably need to be spent collecting in dull, cool weather. If the number of grasshoppers caught in a standard number of sweeps is recorded, the formula in Exercise 1 can be used to decide roughly when enough animals have been captured (use a pocket calculator).

2 As each insect is caught, one student of the pair should hold it gently but firmly and the other should apply a *small* spot of paint on the dorsal surface of the thorax. (If this animal is caught again on subsequent days it will be marked again, so space must be left on the thorax for several more marks.)

3 Place the grasshopper in the sandwich box. All captures by all student pairs should be marked with the same colour on any particular day.

4 When the allocated collecting time has elapsed the marked individuals should be released in the sub-area in which they were caught, the number of adults and nymphs having first been recorded. Release only those not suffering from damage. Record the total marked and the total released.

5 Repeat 1 on the next day (or within the next 2–3 days) using a different paint colour. Mark all grasshoppers caught including any already bearing marks. Repeat this on as many days as is convenient.

6 Calculate the total number of animals in the population on each day (N_i) and its standard error (see Principles section).

7 Having taken the calculation as far as step 11 in the Principles section, it is possible to plot the population trend of the grasshopper over the period of the exercise. Through each plotted population estimate,

draw a vertical line of the appropriate length to represent two standard errors.

Discussion and conclusions

As long as the prerequisites of the methods are met and some measure of their reliability is obtained (e.g. standard error), mark-recapture techniques are valuable. Many factors can influence the magnitude of the error, however. In both the above exercises we cannot be sure that all the relevant assumptions in the introduction were actually justified. Some insects may have been affected by the paint or the handling, even though there may have been no evidence for this on release. Some marks will have been lost, especially in the case of the immature grasshoppers which could moult at least once during the exercise. Their moulting also reminds one that their developmental stage (instar) is not constant and their likelihood of being caught may change accordingly. Mixing of marked and unmarked insects in the field may have been incomplete especially if the weather was cool or wet.

In the laboratory exercises some *Tribolium* adults are probably killed by the sample tubes while others may have escaped from the 'escape-proof' box. However, births do not affect the results of this exercise since only adults are recorded, and there is too little time for recruitment (egg to adults takes approximately 30 days). It is instructive to discuss the relative roles of these and other factors under the actual conditions of the exercises.

In spite of these problems, apparently 'sensible' estimates are often obtained. For instance, Dempster (1971) obtained very good agreement between the estimated number of cinnabar moth pupae in a 60×90m area of heath in April, 1967 (3900 ± 357) and the estimate of adult moths in May (3944 ± 528) using Jolly's formula. He did mark nearly 2000 moths in that year, however, and *c.* 800 were recaptured at least once. Jolly's method is less valuable if only small numbers are marked. Parr, Gaskell and George (1968) used Jolly's method to estimate a grasshopper's numbers. They caught a total of *c.* 130 in a 'small field' and their estimate for N_2, for example, was less than twice its standard error, which provides further evidence for the need to collect large samples.

Further investigations

1 Snails on walls are suitable for a Jolly analysis but because their activity varies enormously with weather conditions (especially rainfall) a longer period than in most insect exercises is necessary, unless the weather is consistently wet and warm. An advantage, however, is that the snails may be given individual painted numbers which provide more information, such as distance travelled, tendency of individuals to settle in conspicuous places etc.

2 Dragonflies are large, conspicuous and fairly easily caught, and are

often used in mark-recapture exercises. Their mobility, however, especially of the immature adults may make it especially difficult to define the population. Immatures of *Ishnura elegans* (a damsel fly), in contrast, often do not leave the parent colony and the species flies at temperatures down to 15°C (Parr, Gaskell and George, 1968), so this species may be the most suitable of the Odonata.

3 Butterflies and moths sometimes form discrete colonies, such as those of the small blue on chalk grassland, but many common species in gardens etc. may be only 'passing through' and do not necessarily represent a relatively static population. The moth *Phyllonorycter messaniella*, however, lays its eggs on the leaves of holm oak (*Quercus ilex*) and samples of adults made at the appropriate time (October–December in Britain) can be considered to represent a relatively static population and thus be analysed by Jolly's method.

4 A simple version of a Lincoln Index exercise could be simulated using beads in a cloth bag. These could be marked as they are withdrawn, replaced and an estimate of the total bead 'population' made.

Section Two

Spatial Pattern

Introduction

Pattern occurs when the spatial arrangement of a species deviates from randomness. It is apparent to all those who examine natural populations that spatial variation is a widespread phenomenon. Several ecologists believe that pattern is a fundamental property of living systems and that its study deserves more attention than it currently receives (Taylor, 1961; Levin, 1976). Since aspects of the physical environment including microclimate, topography and soils are very unlikely to be distributed at random, one can assume that organisms affected by these factors will likewise rarely be randomly distributed. Such influences on species' distribution are termed *extrinsic* factors.

In addition to these there is a range of *intrinsic* factors, i.e. characteristics of the organisms themselves, which impose pattern on their distribution. For plants the intrinsic patterns are caused by factors including chemical inhibition (allelopathy), ineffective seed dispersal and the various mechanisms of vegetative reproduction such as the production of rhizomes or tillers. Animals also possess a multitude of intrinsic characteristics likely to cause spatial pattern. Because animals are mobile these include not only mechanisms of reproduction, like parthenogenesis which might concentrate populations in one area, but a range of behavioural responses which can operate in the direction of aggregation, as in the case of family groups or herding, or in the direction of regularity where, for example, competition for food or space affects the distribution of individuals.

Historically, the origins of the analysis of pattern start with the numerical detection of non-randomness. Greig-Smith (1964) provides a thorough description of the development of pattern analysis and those interested are referred to this work. It suffices here to say that it was quickly established that randomness in nature was found to be the exception rather than the rule, and that the study of the spatial distribution of organisms has since been used to evaluate the factors responsible for the non-random arrangement.

Spatial pattern implies repeatability and the present state of pattern analysis is primarily involved with the identification and numerical description of these repeating units. The patterns considered here are essentially small-scale and at the level of individual organisms. However, there is no reason why the principle involved in the analyses which follow cannot be applied at higher levels of organization, e.g. the population level or at the community level. Single, overall trends across an area due to some environmental gradient are not considered as patterns in this context and are better investigated by regression techniques.

Apart from the academic interest in the growth and reproductive patterns of populations, as studied by pattern analysis, the concepts involved are now being utilized by applied ecologists. For example, the species aggregation characteristics of weeds and insect pests in crops could assist studies aimed at controlling their numbers (see Lewis, 1965; Dean, 1973).

Since pattern is the departure from randomness in spatial arrangements, the first exercises in this section centre on the detection of non-randomness. Continuing from the identification of pattern are exercises which employ measures of the degree of non-randomness and its direction, i.e. towards regularity or towards aggregation. Finally methods which aim to provide more detailed information relating to the scale of pattern are considered.

Unless the aim of a particular investigation centres on the methodology, pattern analysis is rarely a valid end in itself (Greig-Smith, 1961). It is usually the initial stage of a more comprehensive study where the results of the pattern analysis suggest useful lines of further investigation. Most pattern analysis techniques, as with other numerical methods, require repetition before the results can reliably be used to formulate causal hypotheses, which then require testing on new data.

In this section a number of techniques are introduced which have wide applications, yet one must be aware of the fact that each has its inherent strengths and weaknesses, making it more suitable in some situations than others. Current opinion (Usher, 1975) favours an approach which employs a battery of analytical techniques which can provide a more realistic appraisal of the spatial arrangement than any single technique.

Exercises 9 and 10

Taylor's Power Law

Principles

In samples taken from populations where individuals are distributed at random, i.e. independent of one another, the mean number of individuals/sample (m) is equal to the variance (s^2). In natural populations, however, such spatial independence is rare; mutual repulsion may lead to regularity, when the mean will be greater than the variance, which in a perfectly regular distribution is zero. Mutual attraction, on the other hand, or preference for particular parts of the environment, leads to aggregation, when the mean number/sample will be less than the variance; as the density of aggregated populations increases, the proportionate difference between mean and variance also increases. Taylor (1961) found that for a variety of animal species sampled this relationship between variance and mean density obeyed a simple power law:

$$s^2 = am^b$$

This may conveniently be expressed as a regression equation by transforming to logarithms:

$$\log s^2 = \log a + b \log m$$

where log a is the intercept on the vertical axis and b is the regression coefficient. In this equation b is a true population statistic, an 'index of aggregation' describing an intrinsic property of the species concerned in a particular environment; b exhibits a continuous gradation from near-regular ($b \to 0$), through random ($b = 1$), to highly aggregated ($b \to \infty$) (Taylor, 1961; Southwood, 1971).

In the laboratory exercise, two aphid species on broad bean plants are sampled and their indices of aggregation are compared. In the field, the spatial distribution of a common and easily sampled aphid is investigated and experiments carried out to suggest reasons for the observed distribution.

Exercise 9
The aggregation characteristics of two aphid species between the leaves of their host plant

Apparatus

Broad bean plants; aphids; dishes of water with detergent; scissors; leaf area reference sheets.

Preparation

1 The aphid species compared should be the black bean aphid *Aphis fabae* and the pea aphid *Acyrthosiphum pisum*. The former species is the common black aphid found on broad and French beans in summer, and the latter, which is large, green and drops off the plant when disturbed, may be found on peas and sweet peas. If this exercise is carried out in winter, enough aphids to start cultures may possibly be obtained from university or college biology departments, from research stations or insecticide company research laboratories, as they are commonly cultured. In this description, x = the number of students taking part in the exercise.

2 Six to eight weeks before the exercise, sow broad beans (a dwarf variety such as Sutton Dwarf) singly in pots of *c.* 12 cm diameter containing John Innes potting compost No. 2. These plants will be used to culture the aphids; sow x seeds plus 10% to allow for germination failures etc. Keep the pots in a greenhouse (with supplementary lighting and heating in winter).

3 When the plants are about 10 cm high (2–3 weeks later), infest $0.25x$ plants with *A. fabae* and $0.25x$ with *A. pisum* (about 20 aphids/plant in each case). Remove individuals of *A. fabae* from the source plants by prodding individuals with a fine paint brush (size 00) until they withdraw their mouthparts, then pick them up with the splayed bristles of the brush and transfer to the culture plant. *A. pisum* can be knocked off the source plants and handled with a brush. Keep the two cultures as far apart as possible (preferably in separate greenhouses) and use the uninfested plants to maintain the cultures as necessary.

4 At the same time as 3 above, sow $5x$ new beans singly and keep these experimental plants away from the aphid cultures.

5 Two weeks before the exercise, remove all the *A. pisum* by shaking the culture plants in turn over, for example, a white tray. Using a soft-haired paint brush (size 1) divide the aphids among $2x$ plants so that each plant receives 10–20 aphids of mixed sizes. Inoculate a further $2x$ plants with *A. fabae*, but for this species cut up the culture plants into leaf and stem portions and place these, with attached aphids, on the experimental plants.

6 Before the exercise, measure the areas of a range of leaflets from one or two spare bean plants, including some of the small, folded ones at the

top of the plant. Do this by drawing round each one on graph paper, weigh the cut out shape and calculate its area by comparing its weight with that of a square of graph paper of known area. On a sheet of paper, draw outlines of the measured leaflets, and label them with the area so that you have a reference sheet for rough estimates of the areas of the leaflets to be sampled during the exercise. Make copies of the reference sheet (one/2 students) and make them available on the day of the exercise together with some dishes with water and detergent in them.

Procedure

1 Each student is given two plants with colonies of *A. fabae* and two with *A. pisum*. For each plant in turn, remove one leaflet at random, count the aphids (all forms and ages together), and obtain an estimate of the area of the leaflet by referring to the sheet of calculated leaf areas. Leaflet and aphids should be placed in the container of detergent/water when recording is completed, to avoid contamination of other plants. Hold a card beneath the leaflets with pea aphids to catch those that drop off.

2 Repeat this procedure for a total of ten leaflets from each plant; include four of the small, folded leaflets at the plant tip in this total.

3 Calculate density of aphids (nos/10 cm^2 of leaf) for each leaf sampled; calculate mean density and variance for each batch of ten leaves sampled, and enter on the blackboard.

4 Calculate the regression of log variance (y) on log mean (x) for the class data for each aphid species.

5 Compare the regression coefficient for each species with zero, to test for the significance of the relationship, and with 1, to test for significant aggregation or regularity (see Parker, 1979).

6 A regression coefficient which does not differ significantly from 1 only signifies randomness if log a does not differ significantly from 0. To test this, the standard error of the intercept of the regression line on the y axis must be calculated from the following:

$$SE_a = \sqrt{s^2\left[\frac{1}{n}+\frac{\bar{x}^2}{\Sigma x^2}\right]}$$

where s^2 = an estimate of the variance of the y values for any given value of x (see Bailey, 1959; Parker, 1979); $\bar{x}$ = the mean of all the values of x; Σx^2 = the sum of all the x values after each has been squared; n = the number of pairs of observations.

Divide a by its standard error and compare the result with the appropriate value of t in t-tables for $n-2$ degrees of freedom at the 0.05 probability level. If the calculated value exceeds the tabulated value, then log a does differ significantly from 0. If log a is significantly *greater* than 0 we conclude that the organism was aggregated, i.e. log s^2 greater than log $\bar{x}$; if log a is significantly *less* then 0, the organism was regularly spaced between the sample units, even though in both cases b did not differ from 1.

A comparison of the two b values may be made as follows:

$$\text{Calculate} \quad 2\sqrt{\{(SE_{b_1})^2 + (SE_{b_2})^2\}}$$

If the two b values differ by more than the above, they differ significantly; see Bailey (1959) for the calculation of the standard error of b.

Exercise 10
Aggregation in the sycamore aphid and its causes

Apparatus

One 30 cm transparent ruler per pair of students.

Preparation

1 Find enough sycamore (*Acer pseudoplatanus*) trees or saplings with aphids on the leaves so that each pair of students can sample 100–200 of the leaves on the tree allocated to them. The sycamore aphid (*Drepanosiphum platanoides*) is a large, green aphid living on the underside of the leaves; other smaller and darker aphid species occur also but they are much rarer. If there appears to be one or more other species present on more than *c.* 5% of the leaves, select a different tree.

2 Rule columns in the notebook as in 4 below.

Procedure

1 Working in pairs, one student of the pair selects a leaf at random, ignoring the leaves whose lengths are less than *c.* 7 cm. Without handling it, make an estimate of the degree of overlap between the chosen leaf and the leaf (leaves) below it, i.e. estimate whether (*i*) all the lower surface of the leaf could be brushed by others in wind, (*ii*) part of the lower surface could be brushed or (*iii*) there is no overlap and no potential brushing.

2 Count the total number of aphids on the leaf.

3 Measure the length (cm) of the leaf from the apex to the petiole/leaf junction and the maximum width (cm, at right angles to the longitudinal axis).

4 Record the above information in a notebook in the form below, leaving columns *A* and *D* blank.

Degree of Overlap

(*i*) Complete					(*ii*) Part					(*iii*) None				
No. aphids	*L*	*W*	*A*	*D*	No. aphids	*L*	*W*	*A*	*D*	No. aphids	*L*	*W*	*A*	*D*

where L = leaf length (cm), W = leaf width (cm), A = leaf area (cm^2) and D = aphid density (no./10cm^2).

5 Repeat 4 for a total of 100–200 leaves, the students changing roles half way through the sampling.

6 In the laboratory/classroom, calculate the area of each leaf sampled by using the following relationships between sycamore leaf area and length and width; both give a good area estimate but they differ slightly, so, for each leaf, enter the average of the two calculations in the table (Dixon, 1966).

1) Leaf area (cm^2) $= -133.52 + 24.23\,L$

2) Leaf area (cm^2) $= -34.83 + 13.39\,W$

7 Calculate aphid density/10 cm^2 for each leaf:

$$\frac{\text{no. aphids on leaf} \times 10}{\text{leaf area (cm}^2\text{)}}$$

Enter densities in table to two decimal places.

8 For successive 10-leaf batches, calculate mean density ($\bar{x}$) and variance (s^2).

9 Calculate the regression of log variance (y) on log mean (x) together with the standard error of b for all the 10-leaf batches sampled on the tree, i.e. for 200 leaves sampled, the regression consists of 20 pairs of data.

10 Each pair of students now has an index of aggregation for the aphids on their sample tree; these indices may be compared in the same way as those for aphid *species* were compared in Exercise 9.

11 To analyse the data concerning aphid density and leaf overlap, calculate χ^2 on the basis of a null hypothesis that there is no association between leaf status and aphid density. Set out the data in a table prepared as below, entering the number of leaves in each category.

	Degree of overlap			
	Complete	*Part*	*None*	*Totals*
Aphid density 0.5/10 cm^2 or less	74 (47.3)	21 (23.7)	15 (39.1)	110
Aphid density greater than 0.5/10 cm^2	12 (38.7)	22 (19.4)	56 (32.0)	90
Totals	86	43	71	200

12 Calculate the expected (E) values in each category by dividing the product of the two corresponding marginal totals by the grand total. For

example, the expected number of leaves in the low density/complete overlap category is $(110 \times 86)/(200) = 47.3$.

13 Calculate χ^2 using: $\chi^2 = \Sigma \frac{(O-E)^2}{E}$ where O = observed value

$$\text{i.e.} \quad \frac{(74-47.3)^2}{47.3} + \frac{(21-23.7)^2}{23.7} + \ldots$$

$$= 67.00 \text{ with 2 degrees of freedom.}$$

14 Look up the χ^2 value in tables.
In this example, $p < 0.001$ so the data provide strong evidence for association between leaf brushing and aphid density. The principles of χ^2 analysis are described in Parker (1979).

Discussion and conclusions

Comparisons made within or between species on the basis of their aggregation behaviour are useful in population studies. *Aphis fabae* and the peach-potato aphid, *Myzus persicae*, on sugar beet aggregate differently between plants, the latter species having a distribution nearer to random. From this difference, it was inferred that winged individuals of *M. persicae* moved within the crop much more than those of *A. fabae*, perhaps depositing a few nymphs at each settling point. This behavioural difference may explain the dominant role of *M. persicae* in virus transmission within the crop (Watson *et al.*, 1951). Exercise 9 should show that *A. fabae* aggregates between leaves as well as between plants. This species certainly benefits from such aggregations, as aphids feeding in a compact group improve the nutritional quality of the leaf on which they are feeding, with the result that they grow bigger than singly-reared aphids and are more fecund (Dixon and Wratten, 1971).

Other aphids also aggregate on their host plant for nutritional reasons, but rather than improving the quality of a leaf by feeding together, like *A. fabae*, they aggregate because they select leaves of higher nutritional quality than average. On trees, the nutritive status of the leaves can decline so drastically in mid-summer that aphids feeding there suffer a marked reproductive check. On birch (*Betula pubescens*) however, the leaves are not wholly in phase with one another in their nutritional changes, especially in spring and autumn, and the aphids select those leaves of higher than average quality, where they maintain a higher reproductive rate. Their index of aggregation reflects their selection of leaves; it is high in spring and again in autumn (Wratten, 1974). The distribution of the sycamore aphid, on the other hand, seems to be most influenced by physical factors such as exposure to sun and wind (see Fig. 9/10.1) (Dixon and McKay, 1970) and Exercise 10 should confirm that a leaf's position relative to others is at least one factor influencing the aphid's inter-leaf distribution.

To summarize, in natural situations and with sufficient data, true randomness in an animal's distribution is rare. The high frequency with which aggregated distributions occur is a result of a combination of true gregariousness, where the presence of one animal attracts others, and spatial heterogeneity of the environment making some areas inherently more suitable for survival and reproduction. Some of these factors are probably involved in the spatial distribution of *A. fabae* and *A. pisum* on broad bean and *Drepanosiphum* on sycamore.

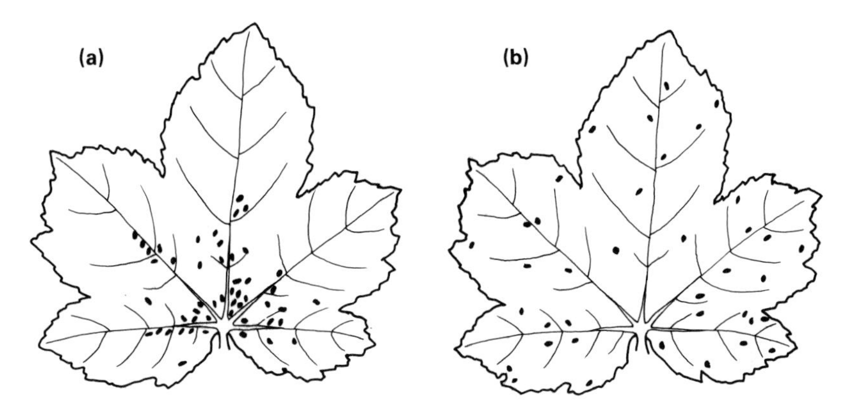

Fig. 9/10.1 The distribution of aphids on a leaf (a) during a high wind and (b) after the wind had abated (after Dixon and McKay, 1970).

Further investigations

1 Any easily-grown plant is suitable for the laboratory exercise as long as two species of aphid (or an aphid and a whitefly, for example) can be cultured on it.

2 Many trees have aphids on their leaves (e.g. oak (*Quercus*), lime (*Tilia*), hazel (*Corylus*), birch (*Betula*)) and the field exercise could easily be carried out on these species, though the reasons for any measured aggregation may not be the same (or as simple) as those applying to the sycamore aphid.

3 The distribution between leaves of the fungal diseases, black spot on roses, and tar spot on sycamore, could also be investigated using Taylor's Power Law.

Exercises 11 and 12

Nearest neighbour techniques

Principles

Techniques based on nearest neighbour measurements of plant or animal distributions on a surface make it possible to describe their pattern in terms of the direction and degree to which the individuals' spacing departs from random.

The basis of the nearest neighbour method, first described by Clark and Evans (1954), is the selection in turn of each individual in the population and the measurement of the distance between it and its nearest neighbour. The mean nearest neighbour distance for all the organisms in a measured area, or a random sample of them, is compared with the mean distance which would be expected if the individuals in a population of that density were randomly distributed. The resulting ratio is a measure of the departure from randomness, the statistical significance of which can be determined. Two or more populations can be compared, using a *t*-test or analysis of variance, respectively, to determine their relative intensity of aggregation. Measuring to the 2nd, 3rd or *n*th nearest neighbour enables the dispersion pattern to be analysed over a larger area and inferences to be drawn concerning the nature of the competition, if any, between individuals or colonies.

Exercise 11
The Distribution of an aphid species on a leaf surface

Apparatus

Single lens reflex camera; monochrome film; photocopier (or Bristol board and transparent book-covering film); dividers and transparent rulers; felt-tip pens.

Preparation

1 On one or more still days from late June to early September, take photographs of the lower surfaces of individual sycamore leaves to include a wide range of sycamore aphid densities (see Preparation section of Exercise 10). The populations should consist mainly of adult (winged) aphids at this time, which simplifies the later analysis. There

will be some white cast skins (exuviae) attached to the leaf from the aphids' last moult. Do not touch the leaf until it has been photographed and try to avoid rapid movement or casting a shadow on it. The aphids have active anti-predator responses and readily jump off the leaf if disturbed. Fill the frame with the leaf and take at least twice as many different photographs as there will be students in the exercise.

2 Measure the maximum length and maximum width of each leaf after it has been photographed, so that its area can be calculated later.

3 Print the photographs life size, using the length and width measurements recorded in the field.

4 Make photocopies of the prints and retain the originals for future use. If the copies are poor or a photocopier is not available, stick each print on Bristol board and cover it with transparent book-covering film, so that having been marked with a felt-tip pen (water-soluble ink) in the practical, they can be cleaned and re-used.

5 On the back of each print or copy, record the leaf area (mm^2); use the equations in Exercise 10, but note that these are in centimetre units.

6 Prepare the blackboard with 3 columns as follows: D, R and d (see below).

7 On the day of the practical, provide each student with two different copies or prints (or only one per student for the highest aphid densities). If not all the prints are used, ensure that the widest possible range of aphid densities is used. Provide also a felt-tip pen and a transparent ruler or pair of dividers per student.

Procedure

1 Proofs of statements below which are marked with an asterisk can be found in the statistical appendix to Clark and Evans' paper.

2 For each aphid in turn, measure (to the nearest millimetre) the distance from the centre of the aphid's thorax to that of its nearest neighbour. Record this measurement (r).

3 Cross out the aphid from which the measurement was made with a short line through its wings, leaving the thorax clear because it may be the nearest neighbour of another individual.

4 Repeat the measurement for each aphid on the leaf, crossing them off in turn. Beware of the presence of exuviae but do not include them in the measurements.

5 Calculate the mean observed distance $\bar{r}_A$

$$\bar{r}_A = \frac{\Sigma r}{N}$$

where N = the total number of aphids on the leaf (i.e. the number of items of data recorded).

6 Calculate the aphid density $D = (N)/(\text{leaf area } (mm^2))$.

7 Calculate the expected mean distance between nearest neighbours

for this density of randomly distributed aphids on a leaf, $\bar{r}_E$

$$\bar{r}_E = \frac{1}{2\sqrt{D}} \quad *$$

8 Calculate the ratio R, which is a measure of the degree to which the observed distribution departs from random

$$R = \frac{\bar{r}_A}{\bar{r}_E}$$

A value of $R = 1$ indicates randomness, $R = 0$ indicates maximum aggregation (all animals on one point) and $R = 2.1491$ indicates maximum possible spacing, i.e. a hexagonal pattern, $\bar{r}_A$ having in this case a value of $1.0746/(\sqrt{D})$*. R therefore has a limited range of from 0 to 2.1491, and in any given distribution on the leaf, the mean observed distance to the nearest neighbour is R times as great as would be expected in a population of randomly distributed aphids of the same density. $R = 0.5$ therefore indicates that the organisms are, on average, half as far apart as expected if distributed randomly.

9 To test the significance of the departure of $\bar{r}_A$ from $\bar{r}_E$, first calculate the standard error of the mean nearest neighbour distance of a randomly distributed population of density D, i.e. $\sigma\bar{r}_E$

$$\sigma\bar{r}_E = \frac{0.26136}{\sqrt{N.D}} \quad *$$

If $\bar{r}_A$ is more than 1.96 standard errors away from $\bar{r}_E$, then the distribution of the aphids on the leaf differs significantly from random ($p < 0.05$, treating $\bar{r}_A$ and $\bar{r}_E$ as normally distributed variables). Other significance levels can be read off from the table of the standardized normal deviate (d) on page 13 in Parker (1979); d in this case is

$$\frac{\bar{r}_E - \bar{r}_A}{\sigma\bar{r}_E}$$

10 Enter on the blackboard for each photograph, values for: density, R and d.

11 Plot the class results for R against density to see if there is a relationship between population density and spatial distribution in the sycamore aphid.

Exercise 12
The distribution of ants' nests, measured by distance to the nearest and to the *n*th nearest neighbour

Apparatus

Bamboo canes; coloured electrical insulating tape; 100 metre tape.

Preparation

1 Find one or more areas of grassland with obvious ant mounds.
2 Using string and canes, divide the areas into working units of known area so that each sub-area contains at least 25 nests.

Procedure

1 Within each sub-area, mark each nest by inserting a cane into the ground at the nest edge.
2 Record the area (m^2) of each sub-area.
3 For each sub-area, prepare one or more note-book pages with columns headed: r_1, r_2, r_3.
4 For each nest in turn, use the tape to measure the distance (in metres to two decimal places) from the *centre* of the nest (not from the cane) to that of its nearest neighbour and enter the distance in the r_1 column.
5 Repeat for the second-nearest neighbour (r_2) and for the third (r_3) for each nest.
6 If for a particular nest, the first, second or third nearest neighbour falls outside the delineated area, delete the records for that nest but leave the cane in position as the nest may be another nest's nearest neighbour.
7 When the three measurements have been made from a nest, mark its cane with the coloured electrical insulation tape to avoid double measurements.
8 In the laboratory/classroom, calculate the density of nests (nests/100m^2) in each sub-area sampled. Call the areas A, B, C etc. and their respective densities D_A, D_B, D_C etc.
9 The measurements should be analysed using the statistic developed by Thomson (1956). He showed that for the nth nearest neighbour, the statistic

$$\Sigma 2\pi D r_n^2$$

is distributed as a χ^2 with $2N_n$ degrees of freedom, where N_n = the number of measurements to the nth nearest neighbours and D is the density. The maximum value which r_n could possibly reach for a particular density will occur when spacing of the individuals is hexagonally arranged (regular). This situation will give the largest value for χ^2, the lowest probability (p) that the value could occur by chance and hence the greatest significance. If the χ^2 value obtained gives a probability of < 0.05 (from χ^2 tables), this indicates significant regularity, whereas if the probability is > 0.95 the individuals are significantly aggregated. A value of p which falls between these extremes indicates that the nests' distribution does not differ significantly from random.
10 For each sub-area in turn, calculate χ^2 for each of the three columns of measurements (r_1, r_2 and r_3) and classify the spacing of the nests with respect to first, second and third nearest neighbours. In other

words arrange a table of results as follows for each of the three classes of measurements (i.e. first, second and third nearest neighbours).

Area	Density	χ^2	df	p	Aggregated, Random or Regular
A	D_A				
B	D_B				
C	D_C				
etc.	etc.				

Discussion and conclusions

Exercise 11 should show a trend, with increasing aphid density, from aggregation to randomness to regularity, and perhaps also a parallel decrease in mean nearest neighbour distance. Although a leaf surface may seem superficially uniform, there is evidence that parts of it are much less suitable for aphid colonisation than others. Perhaps only part of the leaf (such as the tip) may brush other leaves (see Exercise 10), preventing aphids feeding there. Other microclimatic factors may also influence the aphids' distribution, or the reasons for the selection of particular parts may be more subtle; there is some evidence that if aphids feed on very small veins they may cause a blockage in the phloem at the seive plate (Dixon and Logan, 1973). They may also be able to reach the phloem in the large veins less easily because the thicker layer of sclerenchyma cells in these veins impedes penetration by their mouthparts. At low densities, the aphids can select the best sites on the leaf, resulting in aggregation. As density increases, some are forced into less suitable parts and the distribution on the leaf then approaches randomness. Aphids do respond to their neighbours by spacing themselves, however, and at very high densities on a leaf this mutual repellence could lead to regularity.

Evidence for actual competition at some densities and some distances can perhaps be more readily shown by the data from the ants' nests in Exercise 12. It is quite likely that, because of intense intraspecific competition between nests, they have become regularly spaced with respect to their first nearest neighbours. At low densities, this evidence for competition may no longer be detectable for second and third nearest neighbours, their respective χ^2 values probably indicating randomness. At high nest densities, inter-nest competition may extend to second nearest neighbours and this would be suggested by a high χ^2 value for this interval with an associated value of p below 0.05 (see Waloff and Blackith, 1962).

In these two exercises we have been able to show how different degrees of competition between individuals and colonies for favourable parts of a heterogeneous habitat can lead to different spatial arrangements. We can use our classifications of the spacing to deduce how intense the competition has been at different densities.

Nearest-neighbour techniques work well in situations where the individuals or colonies are discrete, easily located and do not move rapidly. The last reason makes Exercise 11 impossible without photography and there are many cases (such as in densely populated plant communities) where it is impossible to decide where one individual ends and another begins.

Further investigations

1 The sycamore aphid exercise could be combined with one based on photographs of aphids taken immediately following strong winds, to see if the density/distribution relationship changes with environmental conditions (see Dixon and McKay, 1970).

2 The familiar red pustules caused by the mite *Eriophyes macrorhynchus* on the upper surface of sycamore leaves could also be used for a nearest neighbour exercise. The leaves can be preserved in alcohol for later use, and nearest neighbour distances measured with dividers. As up to 1000 or more galls can occur on one leaf, it is impractical to include the widest possible density range.

3 Queens of some solitary bees, such as species of *Andrena*, often build their nests at high densities in small areas of grassland or lawn. Each nest is a self-contained underground unit, the surface evidence of which is a small hole with a pile of soil to one side. By careful marking and measurement, information on spacing and competition could be obtained.

4 Molluscs and barnacles on rocky shores are suitable for these methods, as are the burrows and respiratory tubes of annelids and bivalves on sandy shores. Photography could be used in both these situations so they could be used as laboratory exercises.

Exercises 13 and 14

The detection of pattern: comparison with the Poisson distribution

Principles

Techniques based on detection of non-randomness are an important preliminary stage of many ecological experiments. Tests of this type determine whether or not an observed spatial arrangement is random, i.e. has occurred by chance alone. An early approach to the detection of non-randomness was that of Svedberg (1922) and Blackman (1935) who compared the observed numbers of individuals per sample to those expected from a Poisson distribution. This technique has since become a standard test for determining non-randomness in biological populations. The Poisson expectation describes the frequency distribution of a random spatial arrangement, i.e. the expected numbers of samples containing 0, 1, 2, 3, 4...etc. individuals. These theoretical expected frequencies can be compared statistically with those observed from experiments by means of the χ^2 test.

The general arrangement of the data is as below:

No. of individuals per sample unit	0	1	2	3
Observed frequency	X_0	X_1	X_2	X_3
Expected frequency	$n.e^{-m}$	$n.m.e^{-m}$	$\frac{n.m^2 e^{-m}}{2!}$	$\frac{n.m^3 e^{-m}}{3!}$

Where e = base of natural logarithms (2.7183),

m = mean number of individuals per sample unit, estimated by $\bar{x}$,

n = total sample size.

X_0, X_1, X_2 etc. are the observed number of sample units (e.g. quadrats) with 0, 1, 2 individuals respectively.

e^{-m} may be obtained by any one of the following methods:

(1) Refer to Greig-Smith (1964) p. 227;

(2) Refer to the Poisson expected frequencies given in Rohlf and Sokal (1969) p. 215;

(3) As $e^{-m} = 1/e^m$, look up e^m in the table of exponential functions and then find the reciprocal. It is not possible to read off directly every number you may need, so they must be found as follows:

e.g. (*a*) for $e^{4.44}$, find $e^{4.0}$ (=54.598) and then $e^{.44}$ (=1.5527). *Multiply* the two answers to give $e^{4.44} = 84.77$.
(*b*) $e^{4.73}$ may be found for instance by finding $e^{4.0}$ (54.598), $e^{.5}$ (1.6487) and $e^{.23}$ (1.2586) and multiplying the three answers to give $e^{4.73} = 113.29$.

It is the usual practice for the sequence to be halted when any expected frequency class falls below 5, because the χ^2 test is not valid in these cases. For convenience, all observed values beyond this class are summed into a single frequency class. The expected frequency for this class can be obtained by simple subtraction:

$$\begin{matrix}\text{expected frequency} \\ \text{for grouped class}\end{matrix} = \text{total sample size} - \Sigma \begin{matrix}\text{sample units in} \\ \text{all other classes}\end{matrix}$$

Calculate χ^2 as below:

$$\chi^2 = \Sigma \frac{(\text{observed} - \text{expected})^2}{\text{expected}}$$

$$\text{i.e. } \chi^2 = \frac{(X_0 - n.e^{-m})^2}{n.e^{-m}} + \frac{(X_1 - n.m.e^{-m})^2}{n.m.e^{-m}} + \frac{(X_2 - n.m^2 e^{-m}/2!)^2}{n.m^2 e^{-m}/2!}$$

$$+ \text{ etc.}$$

In this calculation the degrees of freedom are 2 less than the number of terms used to calculate the χ^2 value. If this value is greater than the χ^2 value in the table for $p = 0.05$, then the spacing of the organisms sampled differed significantly from random.

Two properties of the Poisson distribution which deserve consideration at the interpretive stage of an investigation are firstly that in a Poisson distribution it is assumed that all organisms are independent of each other and secondly that each site has an equal chance of receiving an organism. When the Poisson expectation is rejected it takes careful experimentation to separate these two factors.

It is possible to derive a useful index of aggregation from the Poisson distribution since the variance/mean ratio is equal to unity in a randomly distributed population. If the calculated value for variance/mean is greater than one, the population has a tendency towards aggregation, and if the ratio is less than one, the population tends towards regularity. The calculation of this simple index extends the use of the Poisson distribution and can assist interpretation. However, a variance/mean ratio equal to unity should not be accepted as conclusive evidence of goodness of fit to a Poisson distribution. It has been shown empirically that a distribution can possess unit variance/mean ratio yet be significantly different from the Poisson model (see Kershaw, 1973, p. 129). A more reliable measure is obtained from the χ^2 test used to compare the observed distribution with that expected from a Poisson.

The significance of the departure of the variance/mean ratio from unity can be tested by calculating $(n-1)s^2/\bar{x}$, which gives a good approximation to χ^2 with $n-1$ degrees of freedom. If the value of χ^2 calculated in this way lies between the values for probabilities of 0.95 and 0.05 (one-tailed test), agreement with a Poisson series is accepted. If the sample is large ($n > 30$), it is assumed that $\sqrt{2\chi^2}$ is distributed normally about $\sqrt{(2v-1)}$ with unit variance. Agreement with a Poisson series is then accepted ($p < 0.05$) if the absolute value of d (i.e. irrespective of sign) is less than 1.96 in

$$d = \sqrt{2\chi^2} - \sqrt{(2v-1)}$$

where d is a normal variable with zero mean and unit standard deviation and v is the number of degrees of freedom. Departures from a Poisson series, and hence from a random distribution, can occur in two directions. The value for d can be:

(1) > 1.96 with a negative sign, indicating a regular distribution;
(2) > 1.96 with a positive sign, indicating an aggregated distribution.

As the variance/mean ratio is influenced by the number of individuals in the sample, it should be supported by the χ^2 test of goodness of fit when samples are large (Elliott, 1977).

Exercise 13
The distribution of flour beetles (*Tribolium* spp.) in their food medium

Apparatus

Wooden or hardboard trays measuring 50 × 50 × 10 cm internally; wholemeal flour to fill trays to a depth of *c.* 5 cm; 100–200 adults of *Tribolium*/tray; single-lens reflex camera with flash; monochrome film; drawing pins (thumb tacks); black thread; plastic teaspoons; 'pooters' (see Preparation section); white tray measuring *c.* 30 cm × 30 cm; large screw-top jar with funnel; PTFE suspension; paper clips; disposable plastic cups with the base removed.

Preparation

The beetles' distribution can be measured either by working with photographs taken at regular intervals or by sampling the colony directly. The two methods measure different aspects of the beetles' behaviour and each method has its own advantages and disadvantages.

Photographic method 1 Several weeks before the photographs are needed, set up one tray with beetles and flour, having treated the inner walls of the box with PTFE to prevent the beetles climbing out (moisten cotton wool pads with the suspension and smear the inner walls; these should be of a smooth material, e.g. the smooth side of hardboard).

2 Place the box on the floor of a room with natural daylight, a

temperature of at least 20°C and no disturbance. Over a period of at least 12 hours in a chosen day, and ideally over the whole 24 hours, take one flash photograph of the flour surface each hour, approaching the box with care and holding the camera vertically above it, filling the viewfinder frame with the box. If an automatic timing facility is available, the camera and flash could be mounted above the colony and set to make hourly exposures over 24 hours.

3 Depending on the number of students taking part, repeat the series on other days to provide one photograph/student.

4 Print the photographs at an enlargement great enough to show the individual beetles clearly, and make sure that the tray dimensions are the same on each print.

5 Prepare transparent sheets (e.g. cellulose acetate or similar material) to fit the prints and using a fine felt-tip pen mark them with a grid of 10×10 squares. The grid should fit the edges of the tray on the photograph exactly. Make as many grids as there are prints.

Colony sampling method 1 Provide one tray of beetles and flour, prepared as above, per pair of students. Using drawing pins and thread divide the tray into a 10×10 grid. Smear the upper inner surface of the jar and the inner walls of the tray with PTFE and provide one tray and one jar for each student pair.

Procedure

Photographic method 1 Taking the squares of the transparent grid as the sample units, count the number of beetles visible in each of the hundred squares and record as a frequency distribution as in the Principles section. If beetles fall exactly on vertical grid lines allocate them alternately to the square to the left or right of the line. For horizontal lines, allocate beetles lying on them alternately to the upper or lower square.

2 Calculate the expected frequencies by using the mean number of beetles/square and the total number of squares (n) in the formula for the Poisson expectation. Calculate also the variance of the 100 counts for each print.

3 Calculate the χ^2 statistic by comparing observed and expected frequencies.

4 Refer to the table of χ^2 with the appropriate number of degrees of freedom.

5 Decide whether or not the beetles aggregated significantly in each one-hour period.

6 Calculate the variance/mean ratio for each hourly print (i.e. $s^2/\bar{x}$) and plot these ratios chronologically to see if there is a detectable diurnal rhythm in the surface activity of *Tribolium*.

Colony sampling method 1 For each square of the grid in turn, insert the bottomless cup, carefully remove the flour and beetles from within the cup with the teaspoon and spread them on the tray. When most of the flour has been removed from a cup and all the remaining beetles are visible, they may be removed with a pooter (Southwood, 1971). When each cup's contents have been counted, tip the tray and pooter contents into the jar.

2 Record the data as before, calculate the expected frequency, the value for χ^2 and the variance/mean ratio.

Exercise 14
The distribution of beetles within a pitfall-trap grid

Apparatus

Wide-necked plastic or glass jars *c.* 10 cm deep and with lids; trowels; sand; bamboo canes; fine forceps; specimen tubes; measuring tape or a metre rule; formalin (40% diluted 1:20 with water) with detergent; key to beetle families (e.g. Lewis and Taylor, 1967).

Preparation

1 Choose an area of rough grassland large enough to take a grid of 8×8 traps at one metre spacings. This is a suitable number for two students to empty and analyse. For more students set up further 8×8 grids. If there is a total of 9 or more student pairs, the grids themselves can be arranged in a larger (e.g. 3×3) grid pattern. The catching period for the traps should be 2–3 days, so if student involvement is to be limited to part of a day, the traps should be set up in advance if the labour is available.

2 Using the trowel, sink a jar with its lid in place in the ground at each point on the 8×8 grid. The top of the jar should be level with the soil surface and there should be no cracks between the jar's rim and the soil. To seal these and make a good junction, trickle sand around the rim and compact it firmly. The lid is to prevent soil and sand falling into the trap during this operation and should now be removed. Mark the position of each jar with a bamboo cane. Pour the formalin and detergent mixture into each trap to a depth of *c.* 2 cm.

3 Decide on a numbering system for the traps.

Procedure

1 When the catching period has elapsed, place a pencil-written numbered label in each trap, replace its lid and take it back to the laboratory/classroom. If it is intended to repeat the exercise, leave the traps in position and remove the beetle catch with forceps and place in numbered specimen tubes.

2 Key the beetle catch to family level. For each family and for all

beetles together in each grid, calculate expected frequencies and compare with observed data as before. Calculate also the variance/mean ratio for each grid.

Discussion and conclusions

The Poisson distribution assumes that each sampling unit offers an equally suitable habitat and that all individuals and sampling units are equal. The final point implies that the expected frequency is not affected by variation in natural sampling units, which is important when leaves, for instance, are the units (see Further Investigations). However sampling units with arbitrary boundaries, as in quadrat sampling, will have expected values related to the size of sample unit. A suitable choice of quadrat size is therefore dependent on a subjective assessment of the vegetation. Those wishing to explore the theory further are referred to Kershaw (1973) and Greig-Smith (1964).

Both these exercises should show that aggregation in animals is very common but that, as suggested in the Principles section, the reasons for the measured aggregation are less easily found. The photographic method reveals an inherent aggregation pattern in the beetles, presumably uninfluenced by the experimenter, but not all the beetles in the population are recorded at one time

The surface aggregations observed are probably related to a diurnal cycle of social behaviour. The beetles are active egg cannibals yet fecundity is related to the number of times a female is mated. The patterns of spatial behaviour observed may therefore relate to some reproductive strategy. *Tribolium* aggregation patterns may also be the result of, or triggered by, an external environmental stimulus such as light or small changes in temperature. Do the results provide any evidence for this? How would one conduct long-term investigations to study which factors may influence the animal's spatial behaviour? Which factors would need to be controlled? (See Mertz, 1972.)

In the field exercise, it is not easy to relate measured aggregation in the beetles to the animals themselves, rather than to some feature of the trapping method. For example, if some traps were sunk deeper than others so that the soil surface sloped down to the jar's lip they might be more efficient. If the aggregation is related to the beetles themselves rather than to the method, then there are many possible reasons. Slope of the ground, plant cover, plant species, nearness to a hedge, localized food could all be involved, but aggregation caused by higher beetle numbers is difficult to separate from that caused by different mobility in different areas (Greenslade, 1964).

Finally, it should be noted that a Poisson series can only be used when the mean number of individuals per quadrat is low relative to the total number that could be accommodated. Where organisms occur at high densities, other distributions, for instance the binomial expansion, are more suitable (Greig-Smith, 1964).

Further investigations

1 Techniques used in the *Tribolium* exercise could easily be applied to the sycamore aphid photographs in Exercise 11. Similarly, the directional pattern exercises (Nos. 17 and 18) could be modified to extract more information from the *Tribolium* and sycamore aphid photographs.

2 An alternative animal to *Tribolium* is the grain weevil, *Sitophilus*, for which wheat untreated with insecticide is used in place of wholemeal flour. Photography may be less appropriate here because there is less contrast between the beetle and the food medium.

3 Many other arthropod groups will be caught in the pitfall traps and χ^2 may be calculated for each one. A temporal element may be included in this exercise by emptying the traps every 2 hours and plotting the variance/meàn ratios chronologically, as in the *Tribolium* photographic exercise.

4 If Longworth small mammal traps are available, a series of nightly catches from a grid of these should be pooled and the aggregation characteristics of each species of mammal calculated.

5 Measurement of the aggregation of woodlice in the field can be made easy by soaking a number of house bricks overnight in water and then placing them (cavity downwards) in a grid on rough grassland. By simply lifting the brick 24 hours later or at intervals during a 12 or 24 hour period, counts and frequency distributions can be obtained.

6 The distribution of the rose leaf fungus disease, black spot (*Diplocarpon rosae*), could be investigated by comparison with the Poisson distribution, either directly in the field or by making a permanent record of a random collection of leaves. This can be done by photocopying leaves which have been stuck on sheets of paper; this works best if the leaves have begun to yellow. The sampling unit is one leaf and the number of spots on each leaf should be recorded. Information on the biology of the fungus and its dispersal can be found in Butler and Jones (1949).

7 A similar investigation to that on rose black spot, above, could be carried out on sycamore leaf-blotch fungus, *Rhytisma acerinum*, and in this case the pattern of infection throughout a woodland could be studied by dividing the woodland into a grid and examining percentage infection. Nearest neighbour techniques (see Exercise 11) could also be used to detect whether diseased trees, or those with a percentage infection greater than an arbitrarily chosen level, occur in aggregations.

8 The use of variance/mean ratio and the χ^2 goodness of fit test for the Poisson comparison could also be used to investigate the spatial arrangement of lawn weeds, using a $1\,\text{m}^2$ quadrat as the sample unit. Species chosen should ideally be well represented over the lawn and individuals should be easily separable. Patterns detected should be related to the known life form of the plants and to their dispersal mechanisms.

9 Distribution studies can be used on a wide variety of vegetation types and the Poisson series has been used to examine, for instance, the regeneration patterns of tree species and the colonization of waste land.

Exercises 15 and 16

Determination of the scale of pattern

Principles

Some of the most difficult patterns to analyse, yet often the most important ecologically, are mosaic patterns. These are patterns where the organism occurs throughout the study area but at different densities. In plant communities the areas of high density are known as clumps and their analysis has provided much useful information in the study of plant competition (Kershaw, 1973) and successional trends (Anderson, 1967). The most frequently used method of analysing mosaic patterns is Greig-Smith's Block-Size Analysis of Variance (Greig-Smith, 1952) which is based on a hierarchical analysis of variance. The technique usually employs a transect of contiguous sampling units (Kershaw, 1957) and any measure of abundance appropriate to the vegetation being studied can be used.

For brevity the technique will be explained using a transect of 16 units (e.g. vegetation records), whereas longer transects (128+) are required for reliable estimates. The calculations only require that the transect units are in the series 2^n to allow the data to be grouped into blocks of 1, 2, 4, 8, 16, 32, 64, 128, etc. The analysis calculates a modified analysis of variance for each block size (BS) by blocking the data as below, in this case using 16 transect units, where a_1, $a_2 \ldots a_{16}$ are the individual measures of abundance of the sampled vegetation.

Data $a_1\ a_2\ a_3\ a_4\ a_5\ a_6\ a_7\ a_8\ a_9\ a_{10}\ a_{11}\ a_{12}\ a_{13}\ a_{14}\ a_{15}\ a_{16}$

Block size 1, $\Sigma x_1^2 = \Sigma(a_1^2 + a_2^2 + a_3^2 \ldots a_{16}^2)$

Block size 2, $\Sigma x_2^2 = \Sigma[(a_1 + a_2)^2 + (a_3 + a_4)^2 + (a_5 + a_6)^2 \ldots (a_{15} + a_{16})^2]$

Block size 4, $\Sigma x_4^2 = \Sigma[(a_1 + a_2 + a_3 + a_4)^2 + (a_5 + a_6 + a_7 + a_8)^2 \ldots (a_{13} + a_{14} + a_{15} + a_{16})^2]$

Block size 8, $\Sigma x_8^2 = \Sigma[(a_1 + a_2 + a_3 \ldots + a_8)^2 + (a_9 + a_{10} + a \ldots a_{16})^2]$

Block size 16, $\Sigma x_{16}^2 = \Sigma(a_1 + a_2 + a_3 + \ldots a_{16})^2$

Sums of squares (SS) for each block size are then calculated by

$$\frac{\Sigma x_1^2}{1} - \frac{\Sigma x_2^2}{2} = \text{SS for block size 1}$$

$$\frac{\Sigma x_2^2}{2} - \frac{\Sigma x_4^2}{4} = \text{SS for block size 2}$$

$$\frac{\Sigma x_4^2}{4} - \frac{\Sigma x_8^2}{8} = \text{SS for block size 4}$$

$$\frac{\Sigma x_8^2}{8} - \frac{\Sigma x_{16}^2}{16} = \text{SS for block size 8}$$

For the largest block size the degrees of freedom (d.f.) $= n-1$ where $n =$ the number of terms in the calculation of the Σx^2 value. The largest BS, which in our example is BS_{16}, has only one term and therefore no SS value can be calculated for this BS. For the next and subsequent block size the d.f. value $= n-1-$(d.f. already accounted for) so that for:

BS_8 d.f. $= 2-1-0 = 1$ i.e. 2^0
BS_4 d.f. $= 4-1-1 = 2$ i.e. 2^1
BS_2 d.f. $= 8-1-3 = 4$ i.e. 2^2
BS_1 d.f. $= 16-1-7 = 8$ i.e. 2^3

Thus the mean square for $BS_1 = SS_1/8$
$BS_2 = SS_2/4$
$BS_4 = SS_4/2$
$BS_8 = SS_8/1$

A graph is then constructed showing mean square against block size (Fig. 15/16.1).

Based principally on the results of ecological investigations, the graphs of mean square/block size have been found to have the following shapes (Brereton, 1971): (*a*) a rising graph with no pronounced steepening or peaks indicates a random arrangement; (*b*) a graph with a low profile exhibiting no peaks, and often parallel with the base, is characteristic of a regular arrangement; (*c*) a graph displaying peaks or pronounced steepening of the profile is characteristic of an aggregated pattern, with the scale of pattern related to the position of the peaks.

Although the applications of this analytical technique have been mainly confined to plant populations there is no reason why animal data should not be analysed in this way. In the following exercises the spatial patterns of heathland plants and of soil arthropods collected from leaf litter are examined.

Exercise 15
The patterns of soil arthropods in leaf litter

Apparatus

A 5 metre length of string; two wooden pegs; 75 × 25 mm specimen tubes; 70% alcohol; ether; Petri dishes; seekers; binocular microscopes; Tullgren funnels; cork borer; wooden dowel to fit inside cork borer; Pasteur pipettes.

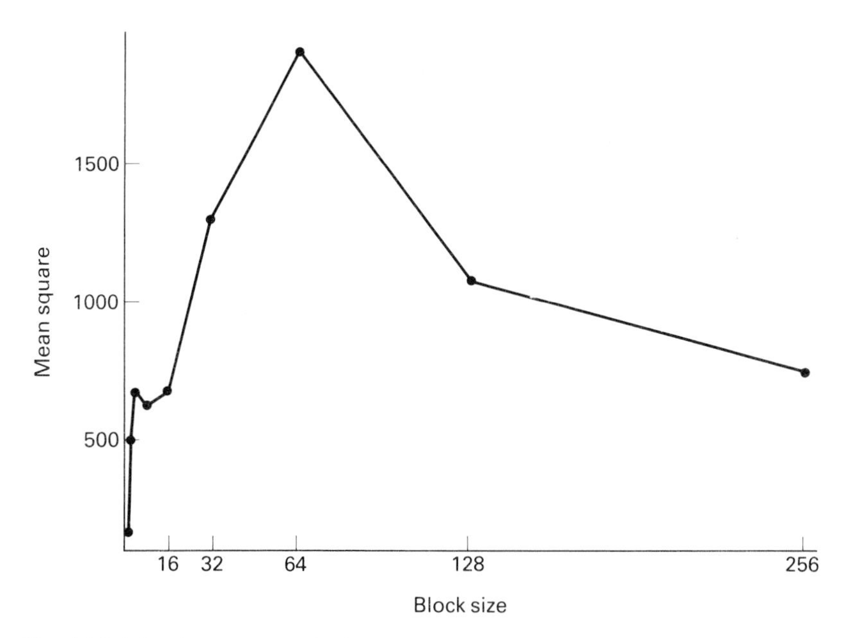

Fig. 15/16.1 A pattern analysis graph (mean square/block size) with a double peak representing scales of pattern at block sizes 4 and 64 (after Kershaw, K. A. (1973). *Quantitative and Dynamic Plant Ecology*, 2nd edition. Edward Arnold, London.).

Preparation

This laboratory exercise requires careful pre-planning, collection of samples and preservation.

1 A site is chosen for the collection of material, which can be carried out at any time of the year. The site should have a well defined leaf litter layer. Oak and mixed woodlands have provided good conditions for this experiment although heathland and parkland have been used.

2 A transect line (the string) is placed along the ground between the two pegs. The sampling procedure is to collect *contiguous* soil samples, i.e. with the minimum possible gap between the holes (128 in this exercise), anywhere along the transect, using the line merely as a straight edge. The samples are collected by pushing a cork borer into the soil to a depth of 50 mm. The actual size of the borer should be recorded since it is this measure that determines the scale of the block-sizes in the analysis. A mark on the side of the cork borer, e.g. a ring of insulating tape, can be used as a depth measure. The size of the cork borer should be selected to fit into the specimen tubes, i.e. < 25 mm, so the soil cores can be transferred easily from the borer to the tubes with a wooden dowel.

N.B. The specimen tubes should be indelibly marked (preferably engraved) from 1–128.

3 On return from the field the samples are treated in a Tullgren funnel

(see Jackson & Raw, 1966). If it is not possible to extract the soil organisms the same day they may be stored in a refrigerator over-night. Small samples as used in this experiment can be extracted using funnels of about 5 cm diameter. Batches of 16 of these can be processed under a single bulb of 25/40 watt *c.* 25 cm above the tubes giving a surface temperature of approximately 30°C. The organisms falling through the funnel are collected in 70% alcohol (in *numbered* specimen tubes) to which a few drops of ether are added to sink the Collembola. The extraction takes only 24–30 hours except with very wet samples.

Procedure

1 Divide the students into groups of 4.

2 The labelled tubes containing the extracted soil organisms are divided among the class so that each student is responsible for several *successive* samples. For instance, a transect of 128 units is divided into sections of 32 units (4 × 8) by a class of 16 students.

3 The numbers of Collembola in each soil extraction are determined by examining the preserved organisms in a Petri dish with the aid of a binocular microscope. Collembola can be readily identified by the presence of the posterior spring (see Chinery, 1973 or Lewis and Taylor, 1967) which is present at all stages of development. A higher powered microscope may be necessary for the identification of some individuals.

4 The data are recorded as;

$a_1, a_2, a_3, a_4, a_5, a_6, a_7, a_8 \ldots$ etc.
where a_1 = No. of Collembola in tube 1 etc.

5 A Block-Size Analysis of Variance is calculated as described on page 61. The data from each group of 4 students are first analysed as separate transects of 32 units giving a total of 4 pattern analyses. Then the first two and last two groups combine their data to give 2 transects of 64 units, and finally the class results are expressed as a single 128 unit transect.

6 Pattern analysis graphs are drawn of mean square against block size and the position of any peaks noted.

Exercise 16
The pattern of *Calluna* heathland in relation to burning

Apparatus

A pin-frame (see Exercise 4) with 2.5 cm between the pins; a tape measure; mallet; wooden stakes 25 mm^2 in cross section.

Preparation

The heathland sites selected for this exercise should provide a wide age range of *Calluna*. Management records help in the locating of such areas and it is often possible to find a mosaic of 4/5 different ages since burning (Fig. 15/16.1) within a few km. Provide students with samples of *Erica* spp. occurring in the same area so that no identification problems arise.

Procedure

1 Students are divided into pairs, one pair to each transect and four transects (minimum) to each site (i.e. age-class of the *Calluna*).

2 Within each age-class the students erect four parallel transects several metres apart with the wooden stakes and tape measures.

3 Along each transect, record for each pin the number of times the *point* touches *Calluna* as it passes down through the vegetation. This is termed cover repetition. The length of the transect should be 256 units (= 6.4 m).

4 Data are recorded as:

Pin 1 2 3 4 5 6 7 8 ...256

No. of hits $a_1\ a_2\ a_3\ a_4\ a_5\ a_6\ a_7\ a_8 \ldots a_{256}$

5 A Block-Size Analysis of Variance is calculated (p. 61) for each transect.

6 Pattern analysis graphs are constructed showing mean square against block size for each transect. Peaks on the graphs are recorded in tables to establish the constancy of their scale:

	Position of Peaks							
Block Size	1	2	4	8	16	32	64	128
Transect 1		√				√		
Transect 2		√			√			
Transect 3			√			√		
Transect 4		√				√		

In this hypothetical example the analyses suggest two scales of clumping, one at BS 2 and another at BS 32 which are equivalent to 5 cm and 80 cm respectively.

7 Compare the patterns of the different ages of *Calluna*.

8 Interpret the patterns by either revisiting the area, if this is close, or by studying large-scale aerial photographs of the area. It is possible to obtain a 1:5 scale photograph by suspending a 35 mm camera from a simple lightweight frame.

Discussion and conclusions

Although the Block-Size Analysis of Variance has not been used widely in the study of animal patterns there is no theoretical reason why it should be limited to plants. In the exercise on soil arthropods the Collembola are nearly always found aggregated, but the scale varies enormously. It is possible to suggest reasons for this behaviour if one considers Collembola in their soil environment (Kevan, 1962). It is probable that since Collembola as a group are weak burrowers (if they burrow at all) they would be restricted to air spaces in the soil caused by the process of leaf fall or more likely by the activities of other soil organisms such as millipedes and earthworms. Reproduction in these limited sites combined with inefficient distribution may also be a cause of local abundance.

So far we have considered passive mechanisms which form aggregations. It is also possible that active attraction to, or avoidance of, some parts of the soil environment operates. Differences in leaf litter composition, the activity of other soil micro-organisms, the distribution of stones and the roots of herbaceous plants may all play a role in the distribution of Collembola.

The temporal perspective is very important in the patterns of soil arthropods. Seasonal changes in their populations are often greater in respect to measures of aggregation than measures of abundance. It would be very interesting to explore the link between the stages in the seasonal break-down of leaf litter and population characteristics of the soil fauna. Smaller aggregations in the winter months may reflect food or microclimate refuges whereas warmer conditions and higher activity levels may explain the larger, less distinct aggregations found during the summer months.

The results of the exercise on *Calluna* heathland is in many ways easier to interpret. *Calluna* when it is not burned severely will regenerate from the base of old burnt stems. This is an immediate source of clumping which can be detected by eye for the first few years. Once the heathland proceeds beyond 5 years since the last burning it usually consists of a mosaic of densities of *Calluna* interspaced with such species as the grass *Molinia caerulea* and the heathers *Erica tetralix* and *Erica cinerea*. This exercise can provide much information about the vegetation structure of heathland at various ages. It is quite usual for two peaks to occur in the pattern analysis graphs which reflect the size of an individual branch and the size of the overall bush. If a smaller sample size was used it would be possible to detect the scale of an individual flower head or leafy stem.

When collected together, the results of several age classes often reveal relationships that could not be readily observed in the field. For instance it is worth considering whether the individual branches increase in size from year to year, or whether it is just their number, or both. Does the mean size of *Calluna* bushes increase annually and if so is there a maximum bush size where heathland is allowed to develop without interference? See Fig. 15/16.1 and the classic work by Watt (1947). A further useful reference is the text on heathland ecology by Gimingham (1975).

Interpretation of the results of pattern analysis can be assisted by reference to past studies. The various scales of pattern encountered in the

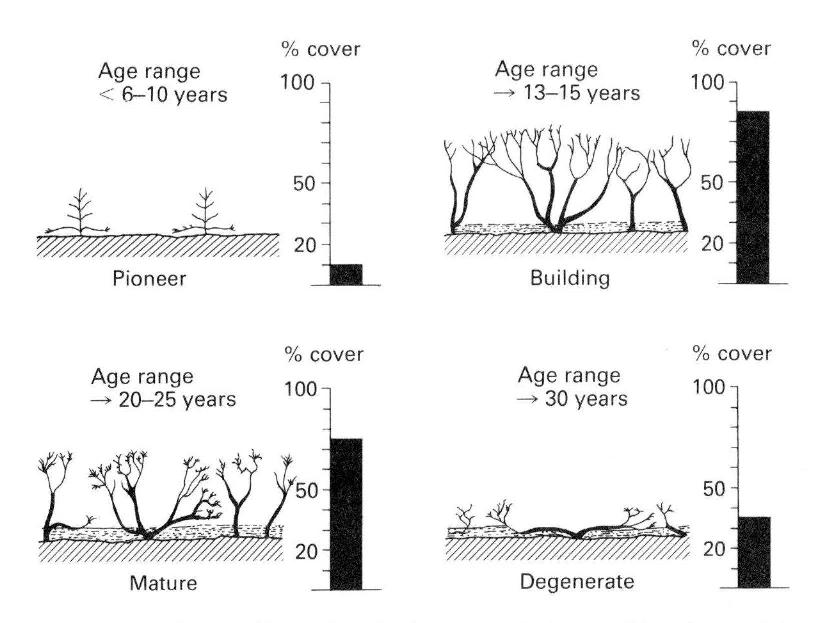

Fig. 15/16.1 Diagram illustrating the four growth phases of heather and the associated change in percentage cover (after Gimingham, 1975).

pattern analysis graphs have been considered (see Shimwell, 1971) as primary, secondary and tertiary depending on whether they are found at the smallest, intermediate or largest block sizes respectively. Ecologically they have been related to small morphologically-determined patterns (e.g. rosettes, tillers or shoots), intermediate sociological patterns resulting from competition between species and large non-repeating patterns such as clonal aggregates caused by large-scale vegetative reproduction or patterns resulting from major environmental changes.

Although many studies have successfully employed Block Size Analysis of Variance, it should be noted that it is an averaging technique and therefore the size of pattern identified is a mean value. This is a potential source of error where two distinct scales of pattern occur within a single block size. Another feature of the technique is its inability to separate clump and gap patterns (Errington, 1973). In its favour, the technique has the attribute of elucidating more than one level of pattern, i.e. it is able to identify situations where the clumps are themselves part of a larger pattern.

Further investigations

1 In the soil extractions, other organisms apart from Collembola will be collected. As the data are readily available it is possible to examine the distribution of other organisms. Those often present in great numbers

are members of the Coleoptera, Thysanura, Hemiptera and Acarina. This study could be further developed by examining the difference in pattern between predators and herbivores/saprovores.

2 A suitable long-term project would be the study of spatial patterns of soil organisms either throughout the year at one site or between sites at one time. The latter would allow comparisons to be made between leaf litter types.

3 Although in this study the spatial distribution was related to small scale patterns, the relationship between numbers of soil organisms (or of a particular group) and the type of leaf litter could be carried out on a large scale. Soil borings could be taken every 10 cm between trees to establish a relationship between soil environment (as measured by organic content, nutrient levels, moisture content etc.) and numbers of organisms.

4 Similarly the collection of soil samples either systematically or at random could be used to investigate relationships and associations between the soil organisms.

5 In Exercises 15 and 16 the possibility of relating the species data to habitat or other species data is simplified by the covariance technique of Kershaw (1961). The statistic involves the calculation, at each block size, of the difference between the expected variance of two sets of data A and B, assuming no relationship, and the observed variance of the two sets of data combined (A + B). The sum of variance A and variance B should equal the calculated variance (A + B) at each block size. When the variance (A + B) value is subtracted from (variance A) + (variance B) the result will be positive for a positive association and negative for a negative association at any block size. A graph displaying these results against block size will possess peaks and troughs at any scales of positive or negative association.

6 Pattern analysis can be used to study the autecology of species, particularly to investigate successional trends in spatial patterning in the seral development of, for example, saltmarsh, sand dune and hydrosere vegetation (see Kershaw, 1973).

Exercises 17 and 18

The detection of directional pattern

Principles

A simple test useful in many ecological investigations examines whether or not a sequence of qualitative records is random. If, for example, the data from a linear grid of contiguous quadrats or some natural series of sample units like the leaves on a twig can be reduced to a sequence of yes/no records, then it is possible to test for randomness. First, the records are scored as + or − depending on the presence or absence of the attribute of interest. For the above sampling methods the records might be the presence/absence of a particular herbaceous plant along a transect through a woodland, or the presence/absence of a species of gall-forming insect on the leaves of an oak tree. In both cases we arrive at a series of records which might look like this:

$$\underbrace{++}_{1}\ \underbrace{--}_{2}\ \underbrace{+}_{3}\ \underbrace{--}_{4}\ \underbrace{++++}_{5}\ \underbrace{---}_{6}\ \underbrace{+++}_{7}\ \underbrace{-}_{8}\ \underbrace{+}_{9}\ \underbrace{-}_{10}$$

The sequence is simply reduced to the total number of runs, i.e. sequences of the same sign. This number and the total number of pluses and minuses are all that is required to test for significance. When the occurrences of a species are tending towards aggregation in the direction of sampling, one would expect fewer runs than that found in a random sequence, whereas if the sequence tended towards regularity, one would expect a higher number of runs than in a random sequence. In cases where neither the total number of pluses nor minuses exceeds 20, there are tables based on the normal distribution which allow both high numbers and low numbers of runs to be tested against values which might occur by chance (see Tables 2A and B).

When long sequences are used it is possible to calculate z-values which are compared with the normal distribution (see Table 1). To calculate a z-score for sequences, the following formula is used:

$$z = \frac{n - \left(\dfrac{2ab}{a+b} + 1\right)}{\sqrt{\dfrac{2ab(2ab-a-b)}{(a+b)^2(a+b-1)}}}$$

where n = the number of runs,
a = the number of pluses,
b = the number of minuses.

The final value will be positive or negative; this does not affect the probability testing (the normal distribution is symmetrical), but does provide additional information. A significant negative value of z tells us that there were too few runs for the sequence to be due to chance, whereas a significant positive value of z tells us that there were too many runs, at whichever probability level one is working (usually the 5% level).

Of the types of data which can be analysed in this way, one of particular interest to ecologists is the possibility of examining directional patterns from a grid. In fact any area data may be used which can be converted to a +/− sequence. Here is a matrix of data records which might be, for instance, mammal trap records in a pasture, or trees in a woodland.

Row values	Sign
4	+
2	−
3	−
3	−
1	−
4	+
4	+
4	+
4	+
1	−
6	+
3	−
3	−
2	−
1	−
5	+
Mean row value: 3.1	

The sign is determined by subtracting the mean row value of the data from each individual row value. Other values including the median, the mode or some preset value of known ecological importance can be employed. A further feature of the dichotomizing process is that any numerical data collected in a contiguous grid can be reduced to a sequence. For instance, the vegetation records from a pin-frame might resemble the matrix above and could be similarly reduced to a sequence, as could records of shoot frequency.

Exercise 17
Linear patterns in coastal vegetation

Apparatus

For each student provide an aerial photograph of a stretch of coastline with a well-developed sand dune system, scale *c.* 1 : 10 000; hand lens.

Preparation

1 This exercise is very rapid and could be used in conjunction with field exercises examining the sand dune ecosystem. These could concentrate on the vegetation changes from the seaward foredunes to the dune slacks on the landward side of the system; see Ranwell (1974).

2 Aerial photographs of most of the coast of Britain are catalogued at:

Air Photographs Unit,
Department of the Environment,
Prince Consort House,
Albert Embankment,
LONDON SE1 7TF

who can supply prints in most cases. Similar catalogue services are now available in many parts of the world—the Remote Sensing Society is a useful contact.

3 The sequence of photographs should cover continuous lengths of coastline. If more than one stretch of coastline is examined then the separate lengths should be coded on the reverse side of the photographs, e.g. coast site 1; picture 4. Where a number of students are working on contiguous photographs, overlap effects can be avoided by marking the overlap directly onto the aerial photograph.

4 In most studies of sand dune systems it is the successional changes from the seaward foredunes to the vegetation of the dune slacks which are investigated. Here the aim is to study the spatial patterns of colonization of one particular zone i.e. the linear colonization patterns along the shore parallel to the sea. It is therefore important to select a site where the main dune-forming species are known and can be identified or delineated on the aerial photographs.

Procedure

1 Each student examines the allocated aerial photograph and takes the first seaward line of dunes colonized (e.g. by *Agropyron junceiforme*

or *Ammophilia arenaria*). In this case we are concerned with the dune hillocks running along the shore parallel to the sea, which will appear as bare or colonized bumps on the aerial photograph.

2 The data are recorded as a sequence of sand dunes (always in the same direction for all students, e.g. East → West) which are either colonized or not colonized. Hence, for each student there will be series of +'s and −'s depicting colonized and uncolonized dunes.

3 For each photograph, the sequence is tested for randomness by using tables or the formula on page 69, as appropriate.

4 For each run of coastline the data from all photographs are grouped to provide a longer sequence which is tested for randomness, using the formula on p. 69.

5 The results of tests are considered, hypotheses postulated and tested later by collecting field data. Alternatively, field records and data on physical aspects of the dune system can be presented to the students who are asked to draw their own conclusions.

Exercise 18
The directional patterns of lakeside vegetation

Apparatus

50 m plastic tape; 50 cm square quadrats.

Preparation

The selection of a suitable site for positioning a transect parallel to the margin of a large lake or reservoir should include a consideration of the species present, accessibility and safety for students, and the nature of the shoreline. For this experiment the following species have been used in the past: *Glyceria maxima*, *Iris pseudacorus*, *Juncus effusus* and *Ranunculus ficaria*. Larger plants like *Phragmites communis* and *Typha latifolia* can also be examined, but a larger quadrat ($1\ m^2+$) would be more suitable. A number of shoreline plants can be examined by allocating students to different species where the vegetation is species rich. Some of the vegetation is dense and it is often more efficient to use a three sided quadrat of wire with a fourth side of twine, so that it can be pushed into vegetation whilst open, then closed when in position. This saves threading the quadrat down through the vegetation.

It is unlikely that the shoreline will be straight, and since the results will be more accurate if the transect is equidistant from the water throughout its length, the transect is therefore likely to be slightly curved. This feature does not create a problem since there is no reason why the transect itself should be straight so long as the data are collected in a continuous series. If the shore is particularly curved it might become difficult to place quadrats without overlapping, and point quadrat samples may be more suitable.

Procedure

1 The aim of the experiment is to establish whether the marginal vegetation of a lake forms sequential patterns along the lake shore. The major step in such an investigation is to test the null hypothesis that the vegetation sequence is random.

2 A 50 m transect is erected along the lake margin to locate 100 × 0.5 m quadrats.

3 The presence or absence of the selected species is recorded as +/− or, in the case of a plant with very high cover values, the shoot density is recorded and the data dichotomized on the mean value (see Principles).

4 The number of runs is counted.

5 The number of pluses and minuses is recorded.

6 The z-value is calculated from the formula on p. 69.

7 The probability that the calculated z-value differs from the expected value assuming a random arrangement of the selected species is obtained from Table 1. Having decided on the sign of the calculated value, a one-tailed test should be used; this tests for the probability of a value being more extreme than z. (To test for the probability of a value being more extreme than either $+z$ or $-z$, use a two-tailed test (Table 1.) The sign of the z-value provides further information; a negative sign indicates a trend towards aggregated units and a positive sign indicates regularity.

8 Interpret the results of each species by referring to its growth form and its reproductive strategy; these may be found in a detailed flora, e.g. Clapham, Tutin and Warburg (1962).

Discussion and conclusions

There are many ecological situations where non-randomness is suspected in one particular direction. The tests described in this pair of exercises allow the rapid examination of sequential records and can be used to guide students to fruitful lines of investigation. In both exercises the method was used to study the patterns of colonization of individual species adapted to a shoreline existence. In Exercise 18 the lake shore was examined by contiguous quadrat sampling to determine spatial heterogeneity at a particular scale, and as with other methods the size of the quadrat should be related to the vegetation studied. For small species 2 cm × 2 cm sample units have been used to detect morphological patterns within an aggregation, and there is no reason why a number of different quadrat sizes should not be used to detect various scales of pattern, as in Exercises 15 and 16. However, in this case it is the larger-scale repeating patterns along the shore that are of interest. If spatial heterogeneity is detected this could be related to substrate, wave action, microtopography or phytosociological factors as well as to intrinsic morphological characteristics.

In Exercise 17 the dunes themselves are employed as natural sampling units to study their initial colonization. Coastal vegetation has been the subject of much research and some of this has concentrated on the mechanisms of primary colonization and subsequent succession. In this exercise, colonization of whole sand dunes is studied on a large scale to investigate whether or not this is an entirely random process. Are fore-dunes, for example, colonized at random or are they colonized in aggregations or regularly along the coast? If they are not random what mechanisms might be responsible for the patterning? Here it is worth considering both spatial heterogeneity in the sand habitat and the intrinsic properties (especially dispersal) of the plant being studied.

In general, one detects little small-scale pattern on newly-colonized dunes, as one might expect from the random arrival of the seed (e.g. wind-blown) or washed-up vegetative propagules. As the density of individuals increases, pattern becomes more intensive, and when the initial colonizing plants become established they form marked aggregations as the dune system is stabilized. The process continues until the aggregations themselves begin to merge, forming a mosaic of different density phases. The causal factors of the pattern may initially be related to an uneven distribution of suitable sites for establishment and growth of the species, and later to patterns imposed by the plants themselves. For coastal species these intrinsic patterns are usually related to rhizome activity, a process which is closely linked to the stabilization of sand and mud substrates. A discussion of the spatial characteristics of colonizing species in a range of coastal habitats is found in Kershaw (1973).

Further investigations

1 Shingle vegetation is another subject suitable for the study of directional patterns. These might be studied at various distances from the high water mark and related to the microtopography of a shingle ridge as well as plant characteristics.

2 The full potential of this method of identifying directional pattern is demonstrated in studies where no obvious environmental gradients or vegetation patterns are apparent. In vegetation mosaics on hummock and hollow or ridge and furrow complexes, for instance, the alignment of the sample grid along and across the area will separate those patterns imposed by the habitat and those originating purely from intrinsic species characteristics.

3 Regeneration of trees in woodlands often reflects spatial pattern which may be in one direction only. The z-test method using transects through the wood will often result in the identification of aggregated patterns. Such patterns are often related to gaps in the woodland canopy or topographic features influencing seed distribution or germination. An investigation comparing seedling distribution data and physical habitat data can reveal relationships that lead to more detailed long-term studies suitable for student projects.

4 Natural vegetation units (cf. the sand dunes in Exercise 17) can be used to compile the sequences needed to analyse directional pattern. For

example, trees infected by disease or parasites can be studied by taking a number of transects at different compass bearings through a woodland and recording the sequences of infected (+) and uninfected (−) trees. Patterns revealed by such analyses may provide evidence of the mechanisms of infection when compared with environmental attributes such as slope, prevailing wind and soil factors, the latter perhaps affecting the trees' susceptibility.

5 If small-scale vegetation maps are constructed for other purposes and contain the positions of shrubs, trees and individual herbaceous plants, it is possible to examine the maps in several directions for directional patterns by this method.

6 On a larger scale, the use of Ordnance Survey maps could enable a study of woods and coppices to be related to topography, major drainage patterns, urban development etc. Similarly, land use maps can be examined for directional patterns, or $2\frac{1}{2}''$ maps can be used for the sequential patterns of parks and green spaces along a transect outward from a city centre.

Section Three

Populations

Introduction

Just as individual animals are born, compete with one another, are preyed upon and die, and plants reproduce, become overcrowded etc., populations exhibit *rates* of reproduction, predation, mortality from competition etc. It is the measurement and interpretation of these rates that concern the population biologist. Studies of the dynamics of population change have obvious value in understanding the reasons for fluctuations in the numbers of weeds or insect pests; such understanding could lead to prediction of pest numbers, manipulation and improvement of natural regulating factors and ultimately to better control of the pest.

Analyses of population structure usually precede attempts to explain population change. The construction of life tables is often the first of such analyses. Time-specific life tables provide information on population mortality at successive time-intervals; age-specific life tables follow the fate of successive 'cohorts' of individuals, such as the eggs, larvae, pupae and adults of a butterfly, and show the level of mortality in each cohort. Since these methods all require a reliable estimate of the number of organisms actually in the population, the exercises in this section also include examples of population estimation methods.

The factors causing the measured mortality, and the importance of these factors in imposing an upper limit on population growth, have long interested biologists, mathematicians and philosophers. Malthus, in his *Essay on Population* (1798) considered that food supply was the ultimate factor which stops population growth. We now suspect that the numbers of many plant and animal species are checked before resources become limiting, and often competition between the individuals within the species is the cause. This can result in the death of some individuals or in a reduction in their size or reproductive rate. In animals, emigration may be another symptom of competition within the species. To regulate the plant or animal population (i.e. impose upper and/or lower limits) these factors must increase their proportionate effect with density. Exercises which measure the effects of competition in plant and animal populations are included in this section, together with some which demonstrate adaptations of populations to environmentally-induced stress.

Exercises 19 and 20

Time-specific life tables

Principles

The time-specific survivorship rate is an important parameter of any population. The numerical value for this characteristic is termed l_x and known as 'the life-table function'. If it were possible to start a study with a large population of new born individuals and record the numbers surviving at the intervals of x_0, x_1, x_2, x_3 etc., then it would be simple to provide the corresponding survival rates l_0, l_1, l_2, l_3 etc. In practice it is difficult to obtain accurate measures of survival, particularly in field situations. Sometimes the ages of animal carcases are used to provide figures, on the basis that the dead animal has lived to the age x but not to $x+1$. Alternatively, a representative sample of the population can be examined and aged. With some animals and woody plants this is possible because they possess annual growth rings, e.g. from teeth of many mammals and scales of fishes (see Taber, 1971). Birds can often be aged from plumage characteristics.

Where a population can be aged indirectly by measurement of size (length or weight), it is possible to construct life tables which reflect the proportion of individuals surviving between size classes rather than time intervals. In many cases the two values will be directly proportional but it is not possible to assume a *linear* relationship. However it is relatively straightforward to provide appropriate regressions for both woody plants and small, short-lived animals.

A life table is constructed on the assumption that the population is stationary, i.e. the age distribution is stable and the size of population constant. A suitable situation which meets this requirement is found in laboratory cultures of insects which have reached their asymptotic (maximum density under the set conditions) population size (see Varley, Gradwell and Hassell, 1973).

To construct a time-specific life table the data are required as age records grouped into arbitrary classes. The numbers of individuals in each class become l_0, l_1, l_2, l_3, l_4 etc. These l_x values relate to the time intervals between classes (which are termed x_0, x_1, x_2, x_3, x_4 etc.) and are measured directly from the population and stated as the number of individuals surviving out of 1000 (this conversion is merely to standardize the numbers for comparative studies and also to avoid working in small fractions). l_4 then becomes the number of individuals surviving at time interval x_4 multiplied by 1000/number of individuals at the start of x_0.

To complete a time-specific life table the following parameters are required:

x = age class.
l_x = number surviving at the beginning of the age interval out of 1000.
d_x = the number dying in the age interval out of the original 1000,
$= l_x - l_{x+1}$.
q_x = the proportion of l_x individuals at the start of the time period that is dead by the end of the period, expressed per 1000 alive at the beginning at the age interval,
$= 1000 d_x / l_x$.
L_x = the number of individuals between age x and $(x+1)$,

$$= \frac{l_x + l_{(x+1)}}{2}.$$

T_x = total number of individuals of age x and beyond,
$= L_x + L_{(x+1)} + L_{(x+2)} \ldots L_{(x+i)}$.
e_x = expectation of life,

$$= \frac{T_x}{l_x}.$$

The interpretation of life tables is usually based on a graphical analysis of the results. l_x is plotted against age; a log/log plot of these values is also constructed; here the slope shows survival rate; q_x and e_x are plotted against age. For all the above if some measure other than age is recorded, e.g. height, then either the values require conversion to age by regression, or the values in the life table may be interpreted in terms of the proportions of individuals surviving between, for example, size classes.

Interpretations of some types of survivorship curves can be found in Slobodkin (1962) and Southwood (1971).

Exercise 19
Survivorship curves for laboratory cultures of *Tribolium confusum* (flour beetles)

Apparatus

A culture of *Tribolium confusum* is used in this example but many insects are suitable if they have overlapping generations and a life expectancy, in the adult stage, of less than 6 months.

Wholemeal flour; culture trays (see Exercise 13), alternatively, large sweet jars, aquaria etc. are suitable; sieve with *c.* 1 mm diam. mesh—nylon flour sieves of this size have been used successfully; quick-drying paint, e.g. cellulose or artists oil paint, of 7 colours; paint brushes size 00; camel hair brushes size 1; photographic developing trays; Petri dishes.

Preparation

1 A *year* before the practical a culture of *Tribolium confusum* is established to provide the stable population required. Many schools and colleges maintain cultures of *Tribolium* for genetics and population studies so a stable population is often at hand. If the culture is set up for the first time a temperature of 30°C and a humidity of 60% will produce a stable population within 3–6 months, depending on the initial culture density. Since *Tribolium confusum* stabilizes its populations at approximately 16 adult beetles per gram of flour, it is possible to calculate the initial numbers required and the total numbers likely to be contained in a set weight of flour. Allow approximately 100 beetles per student in the final culture.

2 *Six months* before the practical (unless the exercise is to be used as a long term project, in which case the students can undertake the following steps), sieve the culture and isolate all the adults. *Tribolium* cannot climb glass or other smooth surfaces, therefore new photographic developing dishes can be used to contain the adults, which rarely fly at room temperature. To make handling easier, the beetles may be cooled as described in Exercise 7.

3 Paint each beetle with a dot of white paint on the thorax. After a few minutes to allow the paint to dry, return the beetles to the culture.

4 Each week on the same day sieve the cultures as before, mark all the new beetles, i.e. unmarked ones, remove the dead adults and return the marked ones to the culture. These are the new recruits to the population. By using a different colour paint each month and coding the weeks of the months by the positioning of the paint, the beetles can be aged to within a week. The beetles may be held quite firmly to mark them, then left in a Petri dish for a minute or two for the paint to harden before returning to the culture.

5 Repeat weekly for 6 months. (If time or labour is limited, 2-week intervals will provide 12 age classes.)

6 After the last marking before the laboratory exercise place the culture in a cool room (*c.* 20°C) to prolong pupal development (thus avoiding the production of new, unmarked beetles). One hour before the practical chill the beetles to make them sluggish and easy to handle.

Procedure

1 Provide students with a detailed schedule listing the stages in the preparatory stage of the experiment.

2 Students work in pairs; one individual to age the beetles, the other to record the data. Students change roles after *c.* 100 beetles have been recorded.

3 Data should be recorded in the form below:

Age (weeks):	0–1	1–2	2–3	3–4	4–5	5–6	6–7	...
Code	Red	Red	Red	Red	Blue	Blue	Blue	
Frequency	f_0	f_1	f_2	f_3	f_4	f_5	f_6	

4 Group class data into a life table using an appropriate age class (in this case, one week).

5 Plot the following survivorship curves for *Tribolium confusum:*
(*a*) l_x against age
(*b*) $\log l_x$ against log age
(*c*) q_x against age
(*d*) e_x against age.

6 Interpret the survivorship curves for adult *Tribolium*. Would these results be expected from natural populations? What other factors might influence mortality?

Exercise 20
Survivorship curves for woodland trees

Apparatus

Per pair of students: tree callipers (or diameter tape); engineering callipers; 25 m tape or measured and knotted (1 m intervals) twine; 4 bamboo canes; string (orange nylon fishing twine is ideal).

Preparation

1 The woodland chosen for this exercise should contain at least one species of tree which is regenerating actively. Many ancient woodlands (200+ years) are relatively stable and thus ideal.

2 Choice of species should be one offering a wide range of sizes at a high density.

3 Familiarize students with seedlings of selected species. N.B. Particular care must be taken when the first pair of leaves (the cotyledons) differ from the normal form as in beech (*Fagus sylvatica*), which exhibits epigeal germination, since this may confound identification.

4 A number of random sites is marked in the wood on the basis that each pair of students can complete $2 \times 25\,\text{m}^2$ quadrats per 3 hour period.

Procedure

1 Allocate each pair of students two $25\,\text{m}^2$ quadrats.

2 Within each quadrat measure: (*a*) the diameter of each tree at breast height (with the callipers); (*b*) the diameter of each sapling, defined as all trees of < 15 cm diameter, at 50 cm above ground, using engineers' callipers; (*c*) the total number of seedlings of < 0.5 cm diameter. Do not record dead trees.

To avoid trampling on unrecorded areas it is important to work through each quadrat systematically. For instance, tape or string can be used to mark lanes within the quadrats, the students recording trees by moving down one lane, back up the next and so on.

3 Each pair of students records the data in frequency classes; 10 cm classes have been found suitable for many tree species.

Diam. of tree	0–10,	11–20,	21–30,	31–40,	41–50,	51–60,	61–70 ... 100+
Frequency	f_0	f_1	f_2	f_3	f_4	f_5	f_6

4 Group the class data into the same frequency classes.

5 Calculate the life table parameters substituting size classes for age classes as below, where n = the no. of age classes:

Size class	No. of living trees	l_x	d_x	*cont.*
$0–10 = x_0$	f_0	1000	$l_0 - l_1$	
$11–20 = x_1$	f_1	$1000 . f_1/f_0$	$l_1 - l_2$	
$21–30 = x_2$	f_2	$1000 . f_2/f_0$	$l_2 - l_3$	
$31–40 = x_3$	f_3	$1000 . f_3/f_0$	$l_3 - l_4$	
etc.				

(Size class)	q_x	L_x	T_x	e_x
x_0	$1000 . d_0/l_0$	$(l_0 + l_1)/2$	$\sum_0^n L_x$	T_0/L_0
x_1	$1000 . d_1/l_1$	$(l_1 + l_2)/2$	$\sum_1^n L_x$	T_1/L_1
x_2	$1000 . d_2/l_2$	$(l_2 + l_3)/2$	$\sum_2^n L_x$	T_2/L_2
x_3	$1000 . d_3/l_3$	$(l_3 + l_4)/2$	$\sum_3^n L_x$	T_3/L_3
etc.				

6 Plot the grouped data as survivorship curves of:
(*a*) l_x/size classes
(*b*) log l_x/log size classes
(*c*) q_x/size classes
(*d*) e_x/size classes.

7 Describe the population trends revealed by the analysis. Which sizes of trees suffer the highest mortality? What do the results tell us about the ecology of woodland regeneration?

Discussion and conclusions

For many animals the survivorship curves obtained in this way have a similar overall shape. This shape reflects a high risk of death amongst young individuals and again amongst elderly individuals, with lower death rates in the middle ages. Caughley (1966) has demonstrated this relationship for a wide range of mammal species. Some animals, in contrast, possess survivorship curves which follow an exponential decline and this is the reason for including the log plot, which would take the form of a straight line in the exponential case. The ecological implications of this form of survivorship curve are that the animal has a constant risk of death (probably through accident or predation) through its life and that negligibly few adults live long enough to suffer physiological ageing. This type of survivorship is common amongst birds (Deevey, 1947).

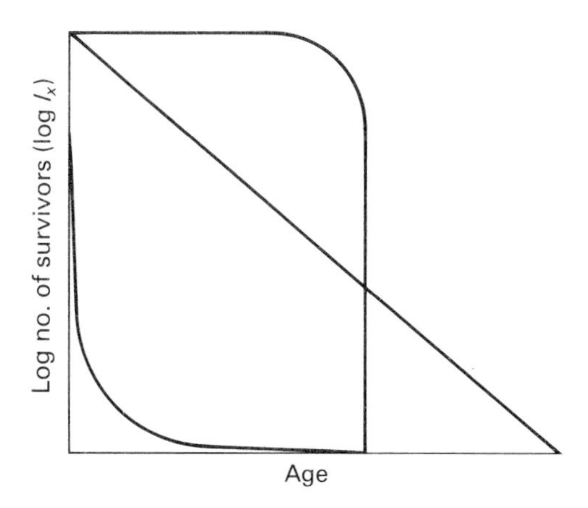

Fig. 19/20.1 The three basic types of survivorship curve (after Dempster, 1975)

A study of the shape of the survivorship curves in these two exercises should enable the student to describe the population trends with age. In the *Tribolium* exercise the final population census provides the detailed age-structure of a culture with a presumed stationary population. Since the beetles only live for between 6–9 months in laboratory conditions there should be a substantial turn-over of individuals in the six month marking period. Which type of survivorship curve does *Tribolium* possess? Are young adult beetles as vulnerable as young mammals?

In the woodland study the survivorship curves reflect the sizes, not the ages, of the trees, but are of particular interest to both foresters and conservationists. It is arguable that for many purposes the structure of a stable woodland is more important than the age of its components. In this exercise the survivorship values relate to the probabilities of a particular size class reaching the next, as well as providing a structural measure of the woodland. What is the nature of the survivorship curve? What does it tell us about natural regeneration cycles? (See also Exercise 23 on self thinning).

Life-tables are a valuable ecological tool for examining the demographic structure of a population; they are capable of summarizing quite enormous data sets and 'bringing order out of chaos'. A classic example of life-table analysis is the work of Miller (1963) who studied the life history of the spruce budworm, an insect of economic importance. In this study life tables were constructed for all the developmental stages of the moth. These are termed 'age-specific' life tables and suitable for the analysis of populations where the generations do not overlap (see Exercise 49). The other feature of age-specific tables is their potential for pest control based on an understanding of the vulnerability of the various stages of an insect's life history (Dempster, 1975).

Further investigations

1 In a study of woodland structure it is possible to obtain direct age measures of seedling populations by counting the ring scars on the main axis. Care should be employed especially with 2–3 year old seedlings where the epigeal germination of many species causes a false ring of a few leaf scale scars. This technique allows the first 5–6 years of growth to be accurately determined but several further age classes can be extrapolated from the lengths of internodal growth between the first, second, third year etc.

2 The relative regeneration states and ages or size structures of various tree species can be compared by recording all the tree species in each of the 25 m^2 quadrats used in Exercise 20 in a mixed woodland.

3 Changing patterns of regeneration may also be investigated in woodlands where an unstable population is thought to exist. If regeneration has recently (< 10 years) been retarded by, for instance, grazing, the life-table analysis will produce some negative d_x values which breaks the underlying mathematical assumptions of the technique but still provides a useful ecological picture, i.e. it tells us the population is not stationary but is declining. The alternative is a population undergoing active growth, as is found in woodlands which are currently being invaded by sycamore, *Acer pseudoplatanus*. Here, the resulting life

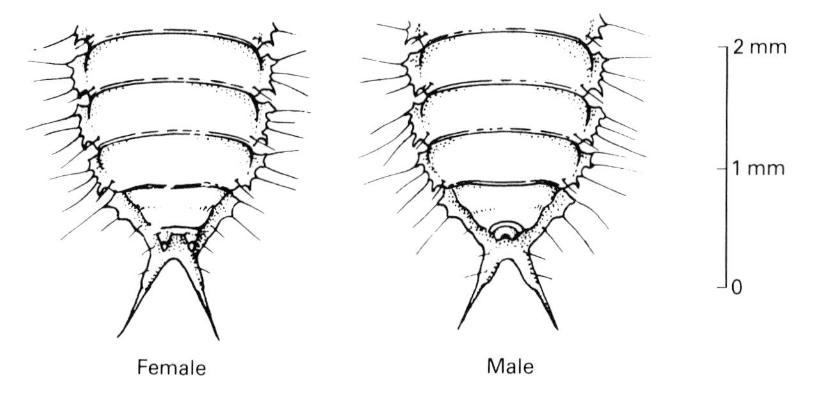

Fig. 19/20.2 Male and female pupae of *Tribolium confusum* (after Darnell, R. M. (1971). *Organism and Environment.* Freeman, San Francisco).

table for this species would show an abnormally steep q_x plot. It should be possible to obtain life tables from numerous woodlands to act as the 'standard' against which other woodlands might be compared. For many tree species the reproductive effort varies from year to year giving mast years with an abundance of seed and other years with little recruitment. What effect would this phenomenon have on the resultant life table? Similarly the colonization of 'gaps' in the forest canopy can be studied. In this case it is often possible to identify the approximate age of the gap and the pattern of colonization.

4 The use of *Tribolium* spp. for population studies is well established and the organism is well suited to this role, being easy to separate from its environment, culture and mark. The egg, larval and pupal stages are also easy to examine and it is possible to study their individual survivorships by separating a single stage and following it through one generation (see Exercise 49).

5 The population growth of *Tribolium confusum* provides a good example of the logistic growth curve. Pupae are relatively simple to sex (Fig. 19/20.2) and can be sorted to provide cultures starting with equal numbers of males and females. These should be started at say 16 (8 pairs) per 16 g of flour at fortnightly intervals and maintained under similar conditions. After 6 months there will be 12 cultures representing different stages of population growth. The adults can be rapidly separated from the flour and counted and their numbers usually give a clear logistic relationship.

6 By visiting a local cemetery and recording the age and year of death and the sex of the individual, time-specific life tables can be constructed for a human population. Divide the data into 5-year age-classes and express the number dead in each age-class as a proportion of the total number of deaths recorded. This, corrected to number dying out of 1000, is then d_x from which values for the other parameters can be calculated.

Exercises 21 and 22

Morphological variation

Principles

In their natural habitats plants and animals show a surprising degree of morphological plasticity. The variation is often in direct response to environmental factors. In plants the variation is quite enormous because of their ability to regenerate. Wind pruned trees are a familiar sight yet few people are aware that most plants will respond to environmental conditions. Heather (*Calluna vulgaris*) for example can take on dwarf forms when severely grazed and bracken (*Pteridium aquilinum*) can be found as dwarf cover 10–20 cm high on exposed heaths yet grow to heights of 200 cm on woodland verges.

Another type of variation occurs due to natural ageing and this produces some of the observed variation in characters such as height and weight found in any population.

In these exercises two examples of morphological variation are considered; the first is the variation between two populations of dogwhelks (*Nucella lapillus*) obtained from different types of rocky coast; the second is the variation in vegetative performance between stands of heather of different ages.

Exercise 21
Ecological adaptations of dogwhelks (*Nucella lapillus*)

Apparatus

Bow callipers with vernier scale; transparent ruler or internal callipers.

Preparation

1 Collect approximately 100 dogwhelks from an exposed rocky shore and 100 from a sheltered rocky shore. Ideal locations in Britain occur off the Welsh coast, Western Ireland, North Cornwall and South Devon. (See Ballentine (1961) for an exposure scale based on biological criteria.) Keep the two samples separate.

2 Boil the dogwhelks in water and remove soft parts with a seeker. These soft parts are not required and may be disposed of.

3 Rinse and dry dogwhelk shells and mark or otherwise ensure the two populations do not mix.

Procedure

1 Measure the length (l) of the shell aperture (excluding the siphon groove) (Fig. 21/22.1).

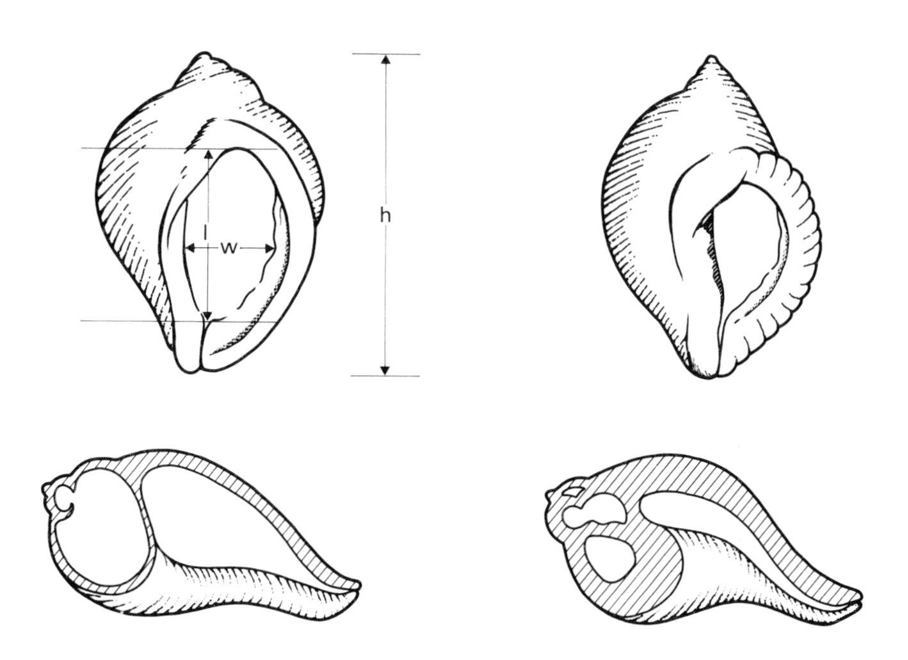

Fig. 21/22.1 Shell aperture shape and thickness in dogwhelks (*Nucella lapillus*) from an exposed shore (left) and a sheltered shore (right). (After Ebling, F. J. *et al.* (1964). *Journal of Animal Ecology*, **33**, 73–82. Blackwell Scientific Publications, Oxford.)

2 Measure the width (w) of the shell aperture at its widest point and measure the height (h) of the shell.

3 Measure the thickness of the dogwhelk shell at a constant point e.g. half way between the tip of the spire and the siphon groove (γ).

4 Calculate the shell aperture index (I_a):

$$I_a = \left[\pi\left(\frac{l+w}{4}\right)^2\right] \Big/ h.$$

6 Tabulate class results for each population of dogwhelks:

Length of aperture	Width of aperture	Shell height	I_a	Shell thickness
l_1	w_1	h_1	$(I_a)_1$	γ_1
l_2	w_2	h_2	$(I_a)_2$	γ_2
etc.				

7 Plot frequency distributions (*c.* 10 classes) for each population using length, width, height, aperture size index and thickness of shell.

8 Calculate *t*-tests between the populations for the measures in 7 above (see Parker, 1979).

9 Plot a graph of shell thickness/aperture size index (γ/I_a) using all the data.

10 Discuss the relationship revealed by this graph and any differences between the populations found in 8 above by referring to the habitats of each population.

Exercise 22
Cyclical variation in *Calluna vulgaris*

Apparatus

1 m^2 quadrats sub-divided into 10 cm squares.

Preparation

As with many of the field exercises, the preparation centres on the selection of suitable sites. In this exercise the requirements are for a range of different ages of heather (*Calluna vulgaris*) (see Exercise 16). To produce sufficient data choose areas close to each other and divide the students equally between areas (a minimum of five areas is required). Alternatively, if the numbers of students are small, two or three contrasting areas can be recorded and additional data for other areas made available during the analysis. Ideally one should aim at providing data for *c.* 5, 10, 15, 20 and 25 years old heathland.

Procedure

Students work in pairs.

1 In each area a number of random quadrats is laid.

2 Within each sub-quadrat the *Calluna* is recorded as present (+) or absent (−) on a > 50% cover basis. The value for each quadrat is found by adding the + values. Thus, if only 20 sub-quadrats contain > 50%

cover of *Calluna* the 1 m^2 quadrat is scored as 20%. As an alternative, pin frame estimates of cover can be used, and although slightly more time consuming give more accurate results (see Exercise 6).

3 Repeat this procedure as many times as possible within each area. In a three hour practical it should be possible to obtain five large quadrats per pair of students. From the management aspect it is easier to operate the practical by having all the students in one area, then to move area *en bloc*, rather than disperse the students, although the latter method provides the greatest number of results.

4 Plot the graph for percentage cover against age as a scatter diagram and draw the best line through the data by eye.

5 Interpret the shape of the graph in relation to the known phases of heather (Watt, 1947; Barclay-Estrup and Gimingham, 1969).

Discussion and conclusions

Morphology is an important aspect of ecology which does not receive sufficient attention in many texts. In these two examples the morphological variation has been well studied and related directly to the overall ecology of the organisms concerned and the ecosystems in which they live.

In the dogwhelk example, the two features recorded, aperture size and shell thickness, although highly correlated are adaptations to different environmental pressures. Kitching, Muntz and Ebling (1966) have been able to demonstrate experimentally that the aperture size variation, which directly relates to the size of the dogwhelk's foot, is an adaptation to wave action. The general explanation is that the larger the foot the greater the power of adhesion. Experiments performed by these workers demonstrated that dogwhelks which were transferred from the sheltered coasts to exposed coasts were unable to remain attached to the rock substrate, presumably owing to the severity of the wave action. Additional investigations showed that in sheltered conditions the population of crabs is higher (in this case the crab *Portunus puber*) and since these are major predators of dogwhelks (they crack the shells with their pincers) this would suggest that shell thickness is an adaptation to such predation (Ebling, Kitching, Muntz & Taylor, 1964). Populations of *Nucella* from exposed shores were substantially more likely to be preyed upon by crabs when transferred to sheltered shores than were the native population.

In the heather practical the cyclical trend demonstrated is one of a rising vegetational cover with age until a certain age after which the cover begins to decline progressively. The sequence shown in Fig. 15/16.1 (p. 67) is well documented by Watt (1955) and Barclay-Estrup and Gimingham (1969) and briefly comprises:

(*a*) The pioneer phase—early growth phase often marked by pyramid-like clumps emanating from the remains of burnt stem stocks. 3–6 years.

(*b*) The building phase—dense hemispherical clumps of heather growing in size until they agglomerate to form a vegetational mat. 10–15 years.

(*c*) The mature phase—the dense clump form of the *Calluna* intensifies until the branches become top heavy and small gaps in the canopy occur. 15–20 years.

(*d*) The degenerate phase—normally the *Calluna* cycle is completed by *c*. 30 years; during the final years the plants become leggy and form a ring of shoots around a central dying plant. 20+ years.

This is only a very approximate guide, for the sequence is continuous and one phase blends into the next. A good summary of the sequence, its effects on other vegetation and changes in biomass of the various components of *Calluna* plants during this cycle are to be found in Gimingham (1975).

Further investigations

1 Dogwhelks from two different shores are compared in this exercise, but a collection from a wide range of habitats could be compared. Wave action can be related to the composition of the shore fauna and flora, obviating the need for the more difficult physical measures. The work of Ballentine (1961) provides a seven point exposure scale based on this premise, suitable for use on the coasts of south-west Britain. The comparison of dogwhelk shell aperture and thickness with points on the exposure scale should reveal interesting trends.

2 Other marine organisms show morphological variation in response to environmental conditions. One of these is the limpet *Patella vulgata*, which exhibits changes in shell height and thickness related to position on the shore. On the upper shore the limpets tend to have thicker shells which are thought to relate to insulation against heat and desiccation. These shells are also taller than lower shore specimens since the organisms are contracted for long periods during shell secretion. On the lower shore, where the limpets are covered for most of the time, the shells are thinner and of a flatter profile.

3 The cyclic patterns exhibited by *Calluna* could be studied by many different sampling techniques if time allowed. Point quadrats used to record cover repetition would provide a measure of changing vegetation 'bulk'. Biomass estimates of the different components of the *Calluna* bushes could be made for the various phases (Freeland, 1970).

4 The changes in flora associated with *Calluna* bushes of different ages is a useful starting point for many projects. In the early phases mosses and lichens cover much of the space between bushes, along with an assortment of flowering herbs and grasses, depending on the type of heathland. As the *Calluna* canopy closes, most of the associated plant species are excluded, but return again during the degenerate phase.

5 *Calluna* bushes of various ages also provide a useful situation for the study of microclimate, which in turn might be used to study the distribution of heathland fauna.

Exercises 23 and 24

Self crowding effects in plants

Principles

Two aspects of self crowding are examined in these exercises. Firstly the phenomenon of self thinning is explored, then the relationship between growth and spacing.

It has long been known (Harper & McNaughton, 1962) that the number of surviving plants beyond a certain density of sowing is not related to the initial seed density. Instead there is a constant relationship between the density of survivors and their total biomass. This relationship has been expressed as a power law by Yoda *et al.* (1963) who proposed the following:

$$W = C\,p^{-3/2} \quad \text{i.e.} \quad \log W = \log C - 1{\cdot}5 \log p$$

where W is the dry weight of the surviving plants, p is the density of the surviving plants, and C is a constant related to the growth characteristics of the particular species being studied.

A number of species have been fitted to this power law by White and Harper (1970). Plots of mean dry weight per surviving plant against density of survivors, expressed as log/log plots, exhibit a gradient of -1.5 within the 5% confidence limits.

The first experiment is an adaptation of the work by White and Harper (1970) and uses rape (*Brassica napus*) to explore the power law.

It is more difficult to obtain field data suitable for analysis since the sequence of events would be difficult to establish in the time available. However it should be possible to establish the power law in woodland situations if a biomass estimate for the trees is available. An alternative, the relationship of tree size to diameter, is given in Exercise 24. Here the relationship between growth and density is looked at in terms of intraspecific competition.

Exercise 23
Self thinning of rape seedlings

Apparatus

20 cm plastic pots; John Innes Compost No. 2; rape (*Brassica napus*) seeds; drying oven; balance; paper bags; pencils; scissors; kitchen towel.

Preparation

1 The experiment is designed to provide three blocks of cultures to cover the age range 5–10–15 weeks. Therefore, for a class of 15, provide 5 repeats of each age range +2 extra of each to allow for failures. This will therefore require a total of 21 potted cultures.

2 Seven plant pots are filled with John Innes compost (to within 2 cm of top), seeds are then evenly sown within each pot at a rate of 150 per pot 15 weeks before the experiment.

3 Repeat 2 at 10 and 5 weeks before the practical and indelibly label the pots.

4 The pot cultures of rape should be grouped close together on a bench in a cool greenhouse and randomized weekly to prevent edge effects. A Latin square technique may be used (see Exercise 30).

Procedure

The data require two periods to collect, which may conveniently be spaced 7 days apart.

1 Cut off the rape plants at ground level and dry them with a kitchen towel if necessary. The aerial parts of the plants are then placed in paper bags.

2 Mark on each paper bag in pencil:

Culture No. ...
Age (weeks) ...
No. of surviving plants ... (density).

3 Place in an oven at 60–70°C.

7 days later

4 Remove paper bags from oven.

5 Weigh the bags containing the rape plants.

6 Weigh the bags empty.

7 Calculate mean dry weight per plant for each culture:
[(wt. of bag + rape plants) – (wt. of bag)]/No. of plants.

8 Plot log mean dry wt. per plant against log density.

9 Calculate the regression of log mean dry wt. on log density (see p. 17).

10 *Note:* The 5% lines either side of the –1.5 gradient have values of –1.25 and –1.83 (White and Harper, 1970). If the slope is within these limits it significantly fits the power law at the 5% level.

Exercise 24
Intraspecific crowding in woodland trees

Apparatus

Tree callipers; 25 m measuring tape; orange electrical insulation tape or yellow chalk.

Preparation

Select a woodland area where one tree species predominates (or where ideally only one species occurs).

Procedure

1 Students are dispersed throughout the study area and work in pairs, one student measuring whilst the other records.

2 Each large tree in the area is taken in turn and its nearest neighbour located. The species of both trees is recorded.

3 The diameter of the tree at breast height and that of its nearest neighbour are recorded, as well as the distance between them. (If a tree has two nearest neighbours select the larger. Note that the measurement of a particular tree does not preclude that tree from being another's nearest neighbour.)

N.B. If callipers are not available, circumference (measured with tapes) may be used throughout this exercise.

4 Mark each tree with a small piece of orange tape or yellow chalk once the diameter/circumference and nearest neighbour measurements have been recorded, to avoid repetition of data.

5 Collect class results.

6 Plot the relationship between the log of the combined diameters of each tree and that of its nearest neighbour against the log of the distance between the two trees for:

(*a*) All pairs of the same species
(*b*) All pairs of mixed species
(*c*) Combined data.

7 Scatter diagrams of this type can be analysed by their correlation coefficients (r). This value is used directly on the log values since these generally approximate to a normal distribution—a requirement of the test.

$$r = \frac{\sum xy - [(\sum x)(\sum y)]/n}{\sqrt{[(\sum x^2 - (\sum x)^2/n)(\sum y^2 - (\sum y)^2/n)]}}$$

where x = log of the combined diameters of nearest neighbours
y = log of the distance between nearest neighbours
n = the number of nearest neighbours recorded.

Calculate r for (*a*) single species pairs
(*b*) mixed species pairs
(*c*) combined data.

8 r can have any value in the range -1 to $+1$ and the significance of r can be found from tables, using $n-2$ degrees of freedom.

9 Interpret the r values as follows:

r not significant: the two values are independent.
r positive and significant: the two variables are directly related.
r negative and significant: the two variables are inversely related.

Discussion and conclusion

Self crowding effects in plants are now known to follow predictable pathways, although the actual mechanism is not fully understood. It is likely, however, that the smallest individuals in a population are the first to die leaving the larger individuals to gain weight. Data from numerous sources and employing species of agricultural or forestry interest comply with the $-3/2$ power law. White and Harper (1970) have shown the work of Bradley *et al.* (1966) on thinning control in British forests to be consistent with the power law, thus demonstrating further the universal application of the law.

Several further relationships may be extracted from Yoda's $-3/2$ power law; given that a plant has a linear dimension of L and covers an area of A then,

$$A \propto L^2$$
$$\text{weight} \quad W \propto L^3$$
$$A \propto W^{2/3}$$

Density (p) is also related to Area (A):

$$A \propto 1/p$$
$$W^{2/3} \propto 1/p$$
$$W^{-2/3} \propto p$$
$$W \propto p^{-3/2}$$

which is the derivation of $W = Cp^{-3/2}$.

Experimentation with the whole range of relationships is possible and a link between spatial pattern (Section 2) and these yield characteristics should be a promising line of future study.

In the woodland study a relationship between tree spacing and tree size is looked for. The hypothesis under test is that there is a positive (direct) relationship between the size of each pair of nearest neighbours and their distance apart. Both of these factors would obviously affect the severity of competition between the individuals. In natural woodlands self thinning would be expected to comply with Yoda's $-3/2$ power law, and larger individuals to benefit from the death of smaller ones. Where regeneration takes place in a woodland gap a large number of seedlings immediately take advantage of the extra light but self thinning operates during their growth to provide only one mature tree to fill the gap. The successful tree then suppresses new seedling growth.

The difference between mixed and same-species pairs introduces another factor, that of estimating the difference between interspecific and intraspecific competition by reference to the corresponding (r) values.

Further investigations

1 As already mentioned, Yoda's power law provides a number of self-thinning criteria related to plant space (volume), density and yield which

could act as the basis for many long term projects. Westoby (1977) believes that Yoda's power law is related to leaf area rather than weight. It should be possible to test this hypothesis for a range of species using a simple leaf area index such as that used in Exercise 10.

2 It is possible to demonstrate the $-3/2$ relationship with a wide range of species and it should be possible to use very quick growing plants, e.g. cress, to enable students to obtain results quickly.

3 Optimal sowing density experiments are ideally suited to analysis by the power law and can be of particular importance when carried out with foliage and root crops. Here, experiments might determine the levels at which thinning operates. At one extreme, density would be so low that self-thinning would not operate and more plants could be grown in the same area (higher yield/m^2), whilst at the other extreme so many seedlings will be thinned as to make the wastage uneconomical.

4 Density in self-thinning experiments eventually reaches an asymptotic level where the surviving individuals have a high probability of growing to maturity and thinning is assumed to be complete. The determination of this density provides other opportunities for ecological project work both in the laboratory and as field investigations.

5 Self-thinning in woodlands may be studied if neglected plantations can be located and yield related to tree girth using forestry tables.

6 The effect of the nearest neighbour was investigated in this exercise but it is possible to relate the growth of a tree to combinations of its 1st, 2nd, 3rd, 4th etc. nearest neighbour. By this method it is possible to determine the sphere of influence of trees upon each other and the number of neighbours influencing each tree (in a closed canopy) (see Exercise 12).

Exercises 25 and 26

Intraspecific competition in aphids

Principles

Direct evidence for intraspecific competition in animal populations (e.g. fighting in vertebrates) is rare, but it may be detected indirectly using circumstantial evidence such as increased death rate, or the individuals suffering a reduction in 'quality' (e.g. size, fecundity etc.) or their leaving the site of competition. In those aphid species whose adults can be winged or wingless, according to the conditions they experience, evidence for competition is often apparent through a reduction in adult size and reproductive rate and through the appearance of winged emigrants in the population. Similarly in aphids whose summer adults are all winged (e.g. some arboreal species), the number of adults which emigrate may increase with increasing population density and presumed competition for food and space. In both cases, the winged adults do not stay to die but, because they are mobile, emigrate. To avoid destruction of the habitat the *rate* of emigration must increase with increasing density (i.e. must be density dependent); sometimes in species feeding on herbaceous plants this rate is not fast enough to avoid killing the plant.

The following two exercises investigate the effects of crowding (i.e. competition for food and space) in aphids and its effects on multiplication rate, production of winged migrants and the rate of emigration.

Exercise 25
Multiplication rate and the production of winged emigrants in relation to density in the black bean aphid, *Aphis fabae*

Apparatus

Broad bean plants; aphid cultures; graph paper; scissors; squares of muslin or terylene net; size 00 paint brush; (optional extras: 1 wood-framed muslin- or terylene-covered cage measuring 100 × 100 × 100 cm with transparent top; binocular microscopes).

Preparation

1 About nine weeks before the exercise, sow 30 broad beans (Sutton

Dwarf or similar variety) singly in 12 cm pots containing John Innes potting compost No. 2. Keep these (watered) in a glasshouse, with supplementary lighting and heating in winter.

2 When these plants are *c.* 10 cm high, infest 10 of them with *c.* 30 aphids each (see Exercise 9) and keep these plants under the same conditions as above but well away from the surplus plants (in the cage or a separate glasshouse if possible). Add some of the surplus plants as necessary as the culture develops. On the same day, sow twice as many new beans as there are students (plus 10%), but a minimum of 36. Keep these well away from the aphid cultures and make sure these plants stay free of aphids; use fingers to squash any colonists.

3 Three weeks before the exercise, infest *half* the newly-sown beans with new-born aphids at different densities. Do this by confining winged aphids (Fig. 25/26.1) from the stock culture on the growing tip (see Exercise 9 for handling details). If insufficient winged aphids are available, use wingless adults (Fig. 25/26.1). Confine the adults on the tip with a square of muslin or terylene net made into a bag by, and held in place with, a *loose* elastic band. The table below gives the approximate number of adults required to give the number of new-born nymphs needed; winged adults are preferable because they nearly always give birth to wingless aphids, whereas wingless adults can produce either form.

Number of colonizing nymphs	*Number of adults required*
2	3
4	3
8	5
16	10
32	15

4 Divide the above treatments equally among half the plants, the sowing programme having provided enough plants for at least three replicates of each colonizing number. Label each pot with the treatment number (i.e. 2, 4, 8 etc. colonists).

5 *c.* 24 hours later, remove the muslin and with a fine (00 size) paint brush remove all the adults and decrease the number of nymphs to a number 2–3 *higher* than the required density. Do not replace the muslin.

6 5–6 days later, remove surplus aphids (all of which should by now be old nymphs, about to moult to adults), to achieve the required density/plant.

7 Two weeks before the exercise, repeat steps 3–5 with the second half of the plants, to give younger colonies on the day of the exercise.

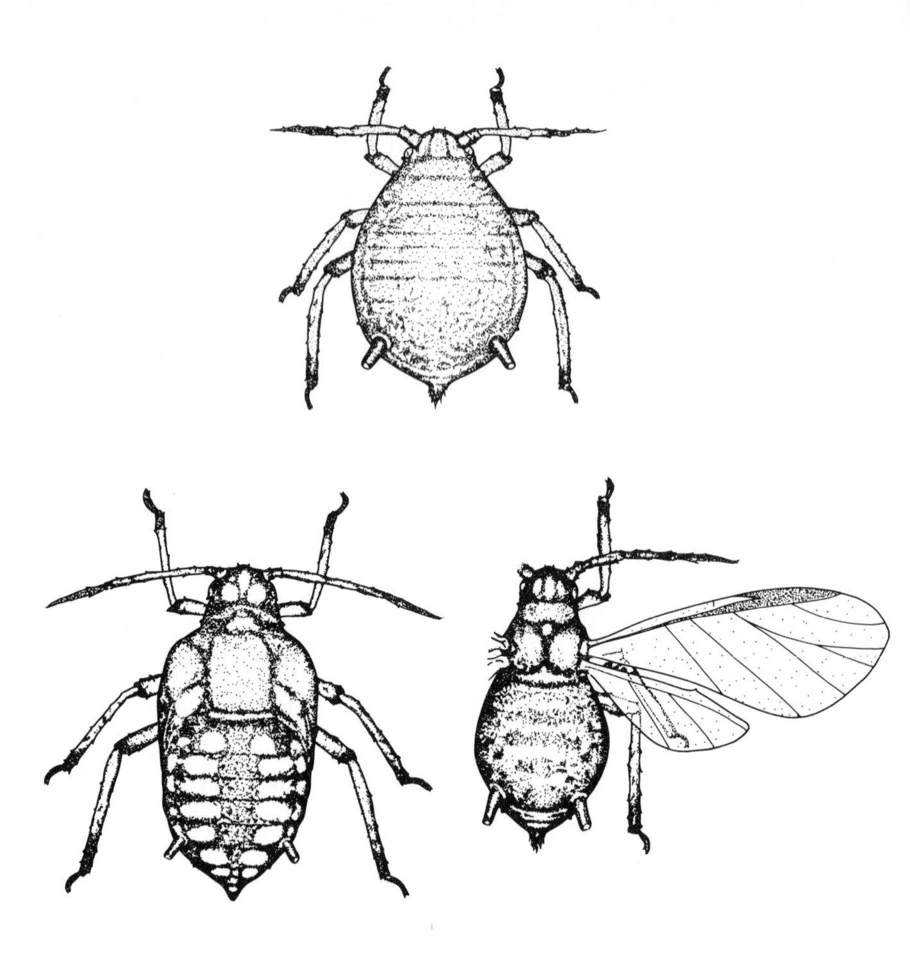

Fig. 25/26.1 A wingless adult, a fourth instar nymph which will be winged as an adult and a winged adult of the black bean aphid, *Aphis fabae*; note the difference in the shape of the cauda ('tail') between the nymph and the adult, which applies to wingless aphids also (after Imms, A. D. (1957). *A General Textbook of Entomology.* Methuen, London).

Procedure

1 On the day of the exercise, divide the 30+ plants among the students so that no individual has more than one high-density (16 or 32 colonists) aphid colony to count.

2 Cut off the leaflets in turn and count the aphids on them, classifying them only into the following classes:

(*a*) winged adults

(*b*) nymphs with wing buds (Fig. 25/26.1); these are easily detected among a colony on a leaf, not by the wing buds themselves but by the two grey swellings on the thorax which are caused by the

developing wing *muscles*. With a little practice, a binocular microscope is unnecessary.
(*c*) others.

3 Count and classify the aphids on the stem in the same way; at high densities the counts will obviously be inaccurate but this is not critical.

4 The class should now have counted at least three replicates for each treatment for each colony age (i.e. 3 early infestations from 2 colonists, 3 late infestations from 2 colonists etc.).

5 Pool the counts for each aphid class for each batch of 3 similar treatments and calculate the mean.

6 Tabulate the means on the blackboard, preparing *two* tables as below, one for the young (late-infested) colonies and one for the old ones. 'Winged' signifies nymphs with wing-buds and winged adults combined.

	Mean No. of aphids counted						
Initial No. of nymphs/ plant	*Winged adults*	*Nymphs with wing-buds*	*Total 'winged'*	*Others*	*Total*	*'Winged' % in total*	*Total/ original colonist*
2							
4							
8							
16							
32							

7 Plot:

(*a*) number of 'winged' aphids against total/plant for young and old colonies in turn. Use the individual plant counts for this, plotting the mean values on the same graph but with a different symbol.
(*b*) % 'winged' aphids in the same way as (*a*).
(*c*) total/original colonist against number of original colonists in the same way as (*a*).
(*d*) % 'others' (young nymphs and wingless adults) against total/plant in the same way as (*a*).

Exercise 26
Flight activity in relation to population density in aphids

Apparatus

Aluminium pie dishes (any shape and *c*. 20 × 20 cm); bright yellow gloss paint; polythene container with *c*. 10 litres of water to which a few ml of detergent have been added.

Preparation and procedure

1 Paint the inside (including the walls) of at least 45 pie dishes with yellow paint.

2 Select a range of at least 15 well-separated (by at least 30 yards) trees with winged aphids on the leaves. Examples are lime (*Tilia*), sycamore, oak, birch, hazel. Select at least 100 leaves from all round the lower canopy of each tree and record with minimum disturbance the total number of aphids (all types) on each leaf. Calculate the mean number/leaf for each tree.

3 Space out on the ground beneath the canopy of each tree 3 yellow traps. Half fill each one with the water and detergent. (The yellow colour attracts flying aphids and the detergent ensures that they drown rather than escape by walking on the surface film.)

4 Return to the traps 24 hours later and count, without removing them, the number of aphids trapped. They should be almost entirely the species from the tree above; do not include in the count individuals which are obviously different from those on the leaves.

5 If the catch is small, the traps can be left in position for several days as long as the aphid populations on the trees are sampled again after 3–4 days.

6 Calculate the mean number of aphids/trap for each tree. Plot this against the mean number/leaf to see if more aphids fly at high densities than at low ones. Calculate the regression coefficient of this relationship with its standard error.

7 Calculate the ratio of mean number/trap: mean number/leaf for each tree and plot this ratio against the mean number/leaf. Examine whether the *proportion* of the aphid population which flies increases with density by calculating the regression coefficient and its standard error for this relationship.

Discussion and conclusions

The results of the *Aphis fabae* experiment should confirm the findings of Way and Banks (1967) who showed that this species reproduced faster (initially) in populations of 8 colonists than in populations of 2, 4, 16 or 32/plant. The reduction in multiplication rate above the eight-colonist level can be attributed to intraspecific competition. It is surprising, however, that this reduction should occur at such a relatively low density. These authors suggested that because this aphid aggregates on its host plant (see Exercise 9), it causes self-induced competition by 'ignoring' apparently suitable parts of the plant. This behaviour, together with an increasing proportion of emigrants at high densities, could regulate the population without destroying the host plant. The tendency to aggregate could also explain why the animals fared better at an initial density of 8 than at densities of 2 or 4; Dixon and Wratten (1971) showed that this species can improve the quality of the leaf on which it feeds if the feeding group is about this size. As a result of the improved food, the aphids are bigger and more fecund as adults than those reared singly. It is not unusual for an animal's peak multiplication rate to occur at intermediate densities rather than at low ones; the usual reason in sexually reproducing individuals is that multiplication rate at the lowest densities may be limited by the inability of the individuals to find mates. This has no role in the 'summer' forms of *Aphis fabae* in Exercise 25 because they reproduce by parthenogenesis.

The above explanation should apply to the two-week-old colonies. In the older (three-week) colonies the higher population levels reached by the day of the exercise have probably led to competition even among individuals derived from the 'optimum' 8 colonists. In these populations, therefore, it is quite likely that the initial infestation density leading to the highest multiplication rate will have dropped to 4 or even 2 (look at the total/original colonist).

A relationship between the number of aphids actually flying and population density should result from the field exercise, but whether a density-dependent relationship can be detected will depend on many factors. In some animals, including arboreal aphids, 'trivial' flight activity (Kennedy, 1961; Southwood, 1962) may be involved, rather than migration. In the former case, the animals do not fly far and may merely be redistributing themselves within the habitat rather than leaving it, as in the latter. One cannot tell whether the trapped aphids were migrating or displaying 'trivial' movement. In either case, however, the movement is likely to have been induced at least partly by population pressure, i.e. competition for food and space (Dixon, 1969).

Further investigations

1 If the size or weight of the adult aphids in Exercise 25 can be determined then the effects of competition on the quality of the individual (in this case size) can also be evaluated.

2 Collections made of aphids from field colonies over a season could be examined to relate the size of the insects on each sampling date to the population size determined at the time of collection.

Exercises 27 and 28

Insect flight

Principles

The physics and physiology of insect flight are complex but some general principles of the mechanics of flight and its ecological interpretation can be derived from simple field and laboratory experiments. The weight carried per unit area of wing (the wing loading), for instance, may be expected to differ between insect species which fly long distances (i.e. migrants) and related species which undergo more local, 'trivial' movements. Whether one would expect migrants to have a higher or lower wing loading than non-migrants depends on which aerodynamic aspects are considered important in long-distance flight. If the weight carried per unit area of wing must be kept to a minimum in migrants one would expect their relative wing loading to be low, whereas considerations of streamlining, wing shape and the reduction of drag might suggest a higher wing loading for migrant species. Exercise 27 compares migrant and non-migrant moth species for evidence of a difference in their wing loading in the light of the above considerations. It is possible to classify moth species as migratory or not by referring to the literature, but the direct measurement of their wing loading obviously requires that they be captured. This is not always easy as they may be local or may fly for only a short period of the year. This problem can be overcome, however, by making measurements from published scale drawings of the migrant and non-migrant species and calculating approximate wing loading from these. Exercise 27 does this for moths and requires only a small preliminary collection of moths from which relationships are established between body volume and weight and between wing area and wing weight. It tests the hypothesis that the wing loading of migrant and non-migrant moth species does not differ.

Because of their abundance and ease of capture, moths also provided early students of insect migration with much data which could be analysed to determine the factors influencing the periodicity of flight and the abundance of the catch. Early hypotheses suggested that insect flight activity increases gradually with increases in temperature up to an optimum, and then declines (Williams, 1940, 1961). When applied to catches comprising large taxonomic groups, positive linear regressions of some measure of activity (corrected for variation in population size) with temperature were obtained. There is little laboratory evidence, however, that over the range of temperatures at which an insect species can fly, flight activity is influenced by temperature. On the contrary, Taylor (1963) suggested and demonstrated that there are upper and lower thresholds for flight for a species, and only between these thresholds will flight occur. To demonstrate these thresholds in the field one could simply relate the number of flying insects captured to the temperature when they were caught. If the response really is a threshold one, rather than a continuous one, and if the population is constant, one might expect no flight below a certain temperature and then similar numbers captured at all temperatures up to the upper threshold. In practice, the aerial population usually does vary with time and hence such a plot would show a scatter of points above the lower threshold. However, one would not expect there to be a significant regression relationship in these

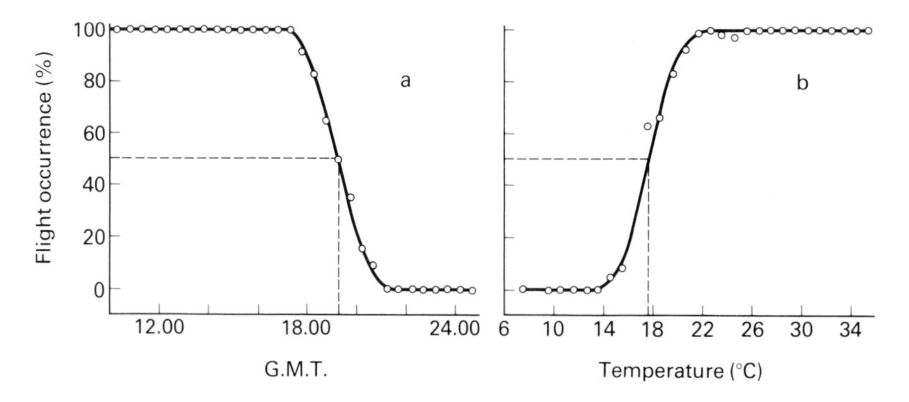

Fig. 27/28.1 Graphical determination of light (left) and temperature (right) thresholds for an insect (after Taylor, 1963).

points if the threshold hypothesis is correct. Variation due to chance or population changes can be largely eliminated if the catches are classified simply as 0 or 1 where 0 = none caught and 1 = one or more captured. The trapping intervals (which could be as long as one day or as short as half an hour) are then grouped according to the measured temperature. The percentage of occasions when flight occurred is calculated for each temperature and these percentages plotted against temperature (Fig. 27/28.1). There is not usually a sharp transition from no flight to 100% flight because of variation in the animals and limitations of temperature measurement, so the threshold for flight can be read off the curve from the 50% flight occurrence point.

Despite the practical problems, field trap catches of individual species do show a relationship like that in Fig. 27/28.1, the accuracy obtained depending on the number of trapping periods which is possible. Exercise 28 is flexible enough to allow for fairly limited trapping time. Analysis of one day's catch data should give a crude indication of the threshold(s) but the exercise is best suited to a residential field course during which the traps can be visited at half-hour intervals on several successive days, the students working on a rota basis.

One problem is that a chosen species may have light intensity thresholds for activity as well as temperature thresholds, and the former may obscure the latter. By emptying the traps only from two hours after dawn to an hour before sunset, we assume that the lower light intensity thresholds are exceeded throughout the trapping period. Because the traps depend on visual attraction, they do not operate at night so that an insect species which flies during the day and the night will not be fully represented in the traps. However, assuming that the range of temperatures encountered during the days when the traps are emptied includes some below the species' lower threshold, a pattern should still emerge.

Exercise 27
Wing loading and migration in moths

Apparatus

As many balances as possible which will weigh items to at least 0.1 mg accuracy; a killing jar (see below). For each student pair: graph paper (mm scale); tracing paper; fine scissors; dividers; ruler; 5–10 dead moths (see below). Have available at least one copy of South (1961)—both volumes, or photocopies (with publishers' permission) of the plates containing those species selected as described below (Preparation 4).

Preparation

1 Make a collection of moths during the summer by removing them from illuminated windows at night or by operating a light trap (see Southwood, 1971). Kill the moths with chloroform or ethyl acetate sprinkled on paper tissues in a screw-top jar (or use crushed laurel leaves instead). Collect 5–10 moths/student pair, but a minimum of 30. Try to collect as wide a size-range as possible.

2 Place the dead, dry moths in a dry container with a lid and some naphthalene to prevent attack by mites. Alternatively, pin them on polystyrene ceiling tiles.

3 A few days before the practical, remove the naphthalene and put the moths in an oven (*c*. 60°C) until needed. This should dry them to minimum weight.

4 Williams (1958) lists moth species which he considers migrants and the text in South (1961) also gives clues to a species' migratory status. Species which are classified as migrants in both texts are given below, together with a representative list of non-migrants.

Migrants	*Non-migrants*
crimson speckled	kentish glory
dark swordgrass	iron prominent
pearly underwing	maple prominent
white-letter wainscot	scarce prominent
white-speck wainscot	plumed prominent
cosmopolitan wainscot	common lutestring
delicate wainscot	oak lutestring
white-point wainscot	pebble hooktip
vines wainscot (rustic)	scarce hooktip
small mottled willow	oak hooktip
dark bordered straw	scalloped hooktip
scarce bordered straw	chinese character
silver Y	scarce vapourer
the ni	square-spotted clay
the alchemist	grey arches
vestal	beautiful brocade
gem	lead-coloured drab
	northern drab.

Procedure

1 Divide the dead moths between the students (5–10/pair).

2 For each moth in turn, and with one student as the measurer and the other as the recorder, measure with dividers the maximum width and length of the body (excluding the antennae) in mm.

3 Calculate the volume of the moth's body on the assumption that it is a cylinder:

$$\frac{W^2L}{4} \text{ where } W = \text{width and } L = \text{length.}$$

4 Cut off the wings as close to their base as possible and weigh them together to the nearest 0.1 mg.

5 Draw round each flattened wing on graph paper, cut out the four shapes accurately and weigh them together.

6 Cut out accurately a known area of graph paper and weigh it.

7 Calculate the moth's wing area (cm^2):

$$\frac{\text{total wt. of 4 wing cut-outs (mg)}}{\text{wt. of 1 cm}^2 \text{ graph paper (mg)}}.$$

8 Calculate the weight of 1 cm^2 of moth wing:

$$\frac{\text{total wing weight (mg)}}{\text{area (cm}^2\text{)}}.$$

9 Weigh the body of the moth (mg).

10 Enter body volume (mm^3) and body wt. (mg) for each moth in a blackboard tabulation, together with the weight of 1 cm^2 of moth wing.

11 While the class proceeds with step 12, the organizer should calculate the equation for the regression of body weight (y) on volume (x); this will be used to calculate the body weight of moths illustrated in South. Alternatively, plot a line by eye and use it to read off body weight in 12 below. Calculate also the mean weight of 1 cm^2 of moth wing.

12 For the migrant and non-migrant species allocated to each student pair, measure, from the illustration, body length and width and calculate the volume. Calculate the body weight when the regression formula is ready or read it off from the line drawn by eye in 11 above.

13 Trace the wings of the illustrated moth, cut out the shapes and calculate the total wing area by comparison with the weight of a known area of tracing paper.

14 Calculate the weight of the wings using the mean value for 1 cm^2 calculated from the dead moths.

15 Add wing weight to body weight to obtain total moth weight.

16 Calculate wing loading for the illustrated moths (mg/cm^2 wing):

$$\frac{\text{Total weight (mg)}}{\text{wing area (cm}^2\text{)}}.$$

17 Enter the moth's name, total weight and wing loading in a second blackboard tabulation, half of which should refer to migrant species and the other half to non-migrants.

18 Plot the class results for wing loading (mg/cm^2) against total weight (mg) separately for migrants and non-migrants on the same graph (use two colours).

19 The plots for the two categories of moths will probably be curved, with wing loading increasing with weight at a declining rate in each case but with the curve for migrants probably *above* that for non-migrants.

20 If the weights are transformed to logarithms, the relationships should become linear and regression coefficients (and intercepts) can be calculated and compared (p. 17), if required. Such analyses will probably not be necessary however, as the difference between the initial curves should be obvious.

Exercise 28
Threshold for flight in insects

Apparatus

For a class of 10–20 students: 16 yellow water traps prepared as described in Exercise 26; a shaded thermometer attached to a 1 m cane (or the use of a nearby Stevenson screen).

Preparation

1 Arrange the traps at least 5 m apart in a 4 × 4 grid on short grass, tarmac, a flat roof or similar site. Arrange the thermometer on the cane near the centre of the grid (if no Stevenson screen is within a few hundred meters), and shade the thermometer with card in such a way that it can be read easily.

2 Pour water with detergent into each trap as quickly as possible. Record the time and temperature when the last trap is filled. Arrange this so that the first student pair can visit the traps half an hour later, which should ideally be in the early to mid afternoon.

Procedure

1 Half an hour after the last trap was filled, the organizer and a pair of students should return to the traps, record the temperature and then visit each trap in turn, taking with them a fresh supply of water and detergent. This and the next few trap visits are concerned mainly with the selection of those insect species which will be recorded on subsequent trips.

2 Walk round the traps, looking for insects which, although probably not identifiable to species, seem to be distinctive and likely to be recognized again. Examples are: 'large, red-eyed muscid fly', 'black aphids', 'thrips', 'orange soldier beetle', 'hoverfly (*Syrphus balteatus*)', etc.

3 Having decided on 5 or 6 suitable 'species', look in the first trap and record presence (1) or absence (0) for each species. A positive record for a species requires only that one individual is found in one of the 16 traps, so once a species is found it can be ignored in all the other traps during that visit.

4 Record all selected species as above. When they have all been found, or the sixteenth trap has been visited, collect a few individuals of each species for laboratory examination.

5 Empty and refill those traps in which at least one of the selected species occurred.

6 Half an hour later, the next student pair should visit the traps with the organizer and again record temperature and presence or absence of the selected species. By this time, one or more of the original choice of 'species' may have to be discarded or modified if laboratory examination reveals that in fact it represented a mixture of species.

7 Empty and refill the traps as before.

8 Continue this procedure for as many days as possible, visiting the traps for the first time two hours after dawn and for the last time one hour before sunset. The organizer should accompany each pair on their first visit to ensure consistency of identification.

9 As each student pair returns to the laboratory, they should record on the blackboard the presence or absence in the traps of each selected species, and the time and temperature. The temperature record should be the average of that visit's reading and that of the preceding one.

10 When the exercise is over, divide the temperature records into

frequency classes of one, two or more degrees. The size of the frequency class will depend on the number of trap visits. Add up the total number of positive records for each species for each temperature class and express this as a percentage of the total number of trap visits made in that temperature class.

11 Plot percentage flight occurrence (% visits with positive records) against temperature class for each of the chosen species. The graph should resemble Fig. 27/28.1b. It may not reach 100% however, because the aerial density of the insects may never have been high enough to ensure a capture in at least one trap, even at temperatures above the flight threshold. If 100% *is* reached, read off the temperature threshold from the 50% flight occurrence point. If 100% flight occurrence is not reached, read off the temperature threshold from the point which is half the height of the peak on the graph.

Discussion and conclusions

To enable wing loading of moths to be determined from scale drawings in Exercise 27 many assumptions had to be made. It was assumed that the weight/volume relationship is the same for migrant and non-migrant species, and that the weight of a unit area of wing is similar between moth species. It could be argued that migrant moths may contain more food reserves (e.g. fat) than non-migrant species, but if this were true, their wing loading would be even higher than that of non-migrants and the graphical separation in Exercise 27 would be even clearer. If we assume that, for a migrant insect, once enough lift has been generated for efficient flight the next priority is reduction of drag, then perhaps the higher wing loading of migrants is ecologically meaningful. A full aerodynamic explanation, however, is doubtless complex and beyond the scope of this book.

Temperature thresholds for insects can be determined crudely by plotting number caught against temperature but the method in Exercise 28 is more precise. In neither case is linear regression the appropriate analysis because of the threshold response. In mixed species populations, demonstrated regression responses (Williams, 1940, 1961) represent the summation of the thresholds of different species at different temperatures and this does give a continuously rising response (Taylor, 1963). The physiological basis of the threshold is probably related to wing-muscle function and the ecological implications are wide, involving especially the distribution of species in time and space.

Further investigations

1 The results derived from illustrations in Exercise 27 could perhaps be verified by using real moths collected in a light trap. It would be useful to compare the weight/volume relationship for migrants and non-migrants.

2 A similar, though no less easily interpretable result may occur if bird species are compared. A book with illustrations of open wings (or measurements of them) would be needed together with data on the birds' weights and their fluctuations.

3 The accuracy of the data in Exercise 28 would be improved if the results from several successive field courses at a site were retained and pooled.

4 Sticky traps (jam-jars inverted on poles and smeared on the outside with horticultural tree-banding grease) could be substituted for the water traps in Exercise 28. These would have the advantage of operating at night as well as during the day (visual attraction is much less important, the insects being caught by more or less random impaction). Removal of the catch is messy and time-consuming however.

5 Regular emptying of a moth trap for several successive nights at half-hour intervals should provide suitable flight occurrence plots for several species.

Exercises 29 and 30

Host plant finding and recognition in insects

Principles

One of the problems facing animals which exploit green plants as food is host-finding. For example, a flying aphid on returning to earth must find one of the handful of plant species on which it can survive and reproduce. Many insects do not specifically identify their host plant from the air but land randomly with respect to plant species. They do, however, show preferences for particular colours: for instance aphids favour long wavelengths, especially yellow, in their landing phase. Since the parts of the plant which are often most favourable for aphid reproduction are the yellowish-green young and old leaves which possess high flow rates of soluble nitrogen, aphids' colour preferences seem to have an ecological advantage. Other insect groups show apparent colour or odour preferences too, but the identification of the character to which they are responding is full of pitfalls. For instance, a colour may be defined by its hue, tint (chroma) and intensity. The hue is the band of wavelengths reflected; the chroma is best defined as the amount of white pigment included in the colours (high chroma represents a low level of white pigment); the intensity is the amount of radiant energy reflected. An insect responding to yellow, therefore, may be influenced by all of these and just because yellow is of a longer wavelength than blue, for instance, does not mean that a higher catch of aphids in a yellow trap is due to a preference for long wavelength light. An added complication is that our appreciation of colour is not necessarily the same as that of an animal. Bees, for instance, respond to wavelengths much further into the ultra-violet than those perceived by man.

In spite of these problems Exercise 30 does attempt to examine the colour preferences of flying insects and by careful examination of the data, some of the difficulties of interpretation can be overcome.

Having found a green plant, the insect must recognize it as edible or not. This usually involves the adult female in finding and recognizing the plant, followed by egg-laying, which implies that immature stages of herbivorous insects do not need to recognize their host, as they are born on it. It is certainly true that the sensory appendages of caterpillars, for instance, are very much reduced compared with those of the adult. Nevertheless, it is quite common for immature insects to find themselves separated from their host plant. They may fall off the leaf, or a colony may consume the whole plant. Exercise 29 should show that lepidopterous larvae do have a food preference and that they can select the plant with which their species has evolved.

Exercise 29
Host recognition by lepidopterous larvae

Apparatus

For each pair of students: 8 dwarf broad bean plants and 8 cabbage plants, all of more or less similar height; 4 Petri dishes with filter paper discs to fit; agar; oven-dried, powdered, cabbage and broad beans (fresh or frozen); *c.* 30 larvae of the diamond-backed moth (*Plutella*) or of the large or small white butterfly (*Pieris* spp.). The former species may possibly be obtained from insecticide company research laboratories or from college biology departments, the latter may be collected from cabbages or Brussels sprouts.

Preparation

1 About 5 weeks before the exercise, sow the cabbage seeds in pairs in 8 cm diameter pots, to provide 8 pots/student pair. If both seeds germinate, remove one seedling.

2 2–3 weeks later, sow the broad beans singly in 12 pots/student pair, to allow for poor germination. Keep both plant species in a glasshouse with supplementary lighting and heating in winter.

3 A week before the exercise, obtain for every 5 student pairs *c.* $\frac{1}{2}$ kg of fresh cabbage and $\frac{1}{2}$ kg of shelled broad beans (fresh or frozen). Chop the cabbage and beans and oven-dry them separately at *c.* 40°C.

4 Grind the dry plant material to a coarse powder in a pestle and mortar or in an electric mill.

5 For each student pair, make up on the day before the exercise *c.* 50 ml of agar solution using 50 ml of distilled water and 1.5 g of agar powder. Heat the agar/water mixture to *c.* 85°C and stir to dissolve the agar. Divide the solution into two batches; add 0.5 g of leaf powder to one and 0.5 g bean powder to the other. Pour each batch into two Petri dishes, fit a circle of filter paper in each lid. Label with the plant contents and allow to cool.

6 Pour off surplus water and dry the agar surface with absorbent paper. Introduce 5 larvae of more or less similar size onto the agar surface, replace the lid and invert the dish. If *Pieris* is used, use medium sized larvae, not the largest.

Procedure

1 Provide each student pair with: 2 inverted dishes with larvae and bean agar and 2 with larvae and cabbage agar; 8 bean plants and 8 cabbage plants; 10 extra larvae.

2 Arrange the 16 plants in a grid on the bench top as below (B = bean, C = cabbage) with the leaves of adjacent plants touching in the rows and columns but not diagonally:

B	C	B	C
C	B	C	B
B	C	B	C
C	B	C	B

3 Add 8 of the extra larvae, using a brush, singly to the plants so that those enclosed by the box in the diagram each receive one larva.

4 Leave the larvae on the plants for at least half an hour, returning to them when the Petri dishes have been dealt with. For each dish, remove the base and count the number of frass pellets (droppings) on the filter paper in the lid. Count the number of dead larvae.

5 Return to the plants and record the position of each larva (cabbage, bean, soil or bench-top) and its behaviour (feeding, walking).

6 Leave the plants in position if possible and return 24 hours later to record the position of the larvae again.

7 For the class results, carry out a *t*-test (Parker, 1979) on:

(*a*) the mean number of frass pellets/bean dish compared with the mean number/cabbage dish;

(*b*) the mean number of larvae alive in each dish category;

(*c*) the mean number of larvae/cabbage plant compared with the mean number/bean. Do this for the 1 hour and 24 hour counts.

Exercise 30
Attraction of flying insects to traps of different colours

Apparatus

For the whole class, provide 36 aluminium pie dishes measuring *c.* 20 × 20 cm and *c.* 3 cm deep. The inner surface of these should be painted (gloss, not emulsion) to give six dishes of each of the following colours: white, black, yellow, green, blue, red. Also required per pair of students: container of water with a few ml of detergent added; fine forceps; 36 corked specimen tubes; paint brush (size 1). At least one light-meter or photographic exposure meter to be shared by the whole class. (Optional extra: infra-red transmitting filter for the light meter.)

Preparation

1 Select an open site of tarmac or mown grass well away from buildings. The best conditions occur on still, warm days from spring to autumn because many of the small insects caught depend heavily on convection currents to remain airborne.

2 24 hours before the exercise, arrange the traps at least 5 meters apart in a Latin Square so that no two traps of the same colour are adjacent in a row or in a column. Numbering the colours from 1–6, one such arrangement is:

1	2	3	4	5	6
2	1	4	3	6	5
6	5	1	2	3	4
4	3	5	6	2	1
3	4	6	5	1	2
5	6	2	1	4	3

3 Pour water with detergent into each trap in turn to a depth of *c.* 2 cm.

4 Record the time.

Procedure

1 Arrange the class so that all traps are visited, giving each trap a number. If students are given help with *in situ* identification of the insect catch in the first few traps, they should be able to record the catch of subsequent traps without taking specimens back to the laboratory.

2 The level of identification depends on the knowledge of the participants and the time available. All specimens should be identified at least to Order; in addition there should be a separate category for aphids. It would also be useful to divide the Diptera (flies) into hoverflies (Syrphidae), Muscidae, Calliphoridae and as many further families as possible.

3 When the individuals cannot be identified to the level required *in situ*, a specimen tube should be dipped into the water in the tray and half-filled. The individual specimens are then transferred to the tube using forceps or a paint brush. A pencil-written label stating trap number and colour is then added and the cork replaced.

4 When the catch of all the traps has been removed or identified *in situ*, remove and discard the larger insects such as large flies still in the traps. Then take a reading of the intensity of the light reflected from each trap in turn by holding the meter at a standard height (e.g. *c.* 20 cm) above each one. If an infra-red filter is available, repeat these readings with the filter in position.

5 In the laboratory/classroom, carry out an analysis of variance for each insect group in turn for the combined class results. Decide whether there is significant heterogeneity between the numbers caught/colour for each insect group (i.e. by considering the significance of the F-value) and compare particular pairs of colours as appropriate using catch totals with their standard errors.

6 Rank the 6 colours according to (*a*) their attractiveness to each

insect group in turn (mean number caught/trap), (*b*) the mean intensity of reflected light and (*c*) the portion of the measured reflected light which is in the 'infra-red' part of the spectrum, as defined by the characteristics of the filter used. In (*c*), this proportion is:

$$\frac{\text{reading with filter}}{\text{reading without filter}} \times \frac{100}{1}.$$

7 Compare the rankings in 6 above and look for those insect groups where the numbers caught do not follow exactly the trend in measured intensity of the trap colours or in the infra-red proportion. Cases where an insect group is attracted most to a colour the intensity of which is not the highest of those measured suggest a definite hue preference.

Discussion and conclusions

The laboratory exercise should demonstrate a food preference even among immature insects with reduced sense organs and this should reflect their feeding rate and survival on the agar mixtures. Insect larvae can discriminate between experimental foods of different nutrient levels as well as flavours, and even between different nutrient combinations (House, 1969; Schoonhoven, 1972).

Many insect groups should show an apparent colour preference in the field exercise, the two commonest ones being those of aphids for yellow and thrips (Thysanoptera) for white. Thrips are often flower feeders so a white preference may be related to this. However, white paint probably reflects more ultra-violet than the other colours used, its intensity is probably the highest and its reflection of infra-red may also be high. Which of these most influences the thrip catch is difficult to decide. Other insect colour preferences seem to make ecological sense, even if we cannot be sure of the nature of the attraction. Many fly species are saprophagous or fungus-feeders and it is often the darker traps which attract this group; it is tempting to suggest that this is related to the flies' tendency to seek out dark, damp places for oviposition.

Further investigations

1 Some insects, like the caterpillars in Exercise 29, recognize a plant as suitable not by its nutritional quality alone but by the presence of characteristic 'odd substances' which define the plant group but seem to have no metabolic role within the plant. The incorporation of the chemical sinigrin (a common constituent of the Cruciferae) at the rate of 0.025 g/25 ml of the agar bean mixture should increase the number of frass pellets counted. This suggests that bean is not repellent to these cabbage-feeding caterpillars but simply lacks the correct feeding stimulant. A fourth treatment of powdered tomato or potato leaves (which *do* contain a repellent) should provide interesting results. By

putting different agar mixtures in the compartments of a 'Repli-dish', a choice chamber for larvae can be constructed.

2 In the field, if only yellow traps are used, they can be placed at 1 meter intervals and at 90 degrees to a hedge or fence which should be more or less at right angles to the prevailing wind. Ten traps on either side of the fence is sufficient. The number of insects caught in each trap should reflect the greater settling rate near the barrier, due to turbulence. Examination of the size distribution of the trapped insects in relation to distance from the barrier could be related to their powers of flight and their ability to respond to the trap when the wind speed is high, i.e. away from the barrier (see Lewis, 1965).

3 Some insects, especially aphids, respond to the *contrast* between a colour and its background. Using yellow traps placed on tarmac, on grass, on bare soil and on a support above a crop this effect could be quantified (see Dempster and Coaker, 1974).

4 By using a wide size-range of yellow water traps, the catch/unit area could be calculated for each one and plotted against the trap area. If the relationship is non-linear, the units on the x axis could be transformed to logarithms or square roots and an ideal trap-size calculated.

5 'Day-glow' paints which differ from normal ones in their ability to convert ultra violet to visible light, could be included in the range of trap colours used in Exercise 30.

6 Vertical baffles, dividing the catch into 90° quadrants, give information on the proportion of the catch derived from each direction and the degree of dependence of each insect group on the wind.

7 If the traps in Exercise 30 are emptied every two hours from dawn to dusk for one or more days, information on the diurnal periodicity in the flight of the major insect groups is obtained. A series of daily catches over several weeks could be related to weather conditions such as wind speed, temperature and humidity. A multiple regression programme in a computer could show which environmental factor accounts for most of the catch variation between days.

8 A sophisticated but expensive way of overcoming some of the interpretative problems of using a range of hues of differing intensities would be to design a standard electrical light source shining through a water trap. Above the trap, neutral-density filters could be added or removed to standardize the intensity of all the traps so that differences in catches could be attributed to hue preferences with more confidence.

9 The preference of landing aphids for long wavelength light has been exploited occasionally in crop protection by placing bands of aluminium foil between the plant rows. This reflects the short wavelength light from the sky and as a result many aphids do not land but fly away. By placing yellow traps in the centre of foil strips on the ground, catches of aphids could be compared with those from yellow traps on bare ground or grass.

Section Four

Population interactions

Introduction

The population interactions dealt with in this section are interspecific competition, allelopathy and predation. Insect parasitism, which is really a form of predation in which one parasite larva kills one host, is included in a community analysis exercise in the next section. Symbiosis, commensalism etc. are not included.

One of the many attempts to define competition was that of Birch (1957). He said that it occurs when a number of organisms (of the same or different species) utilize common resources that are in short supply (exploitation competition) or at least that the organisms harm one another in the process of trying to obtain the resource (interference competition). Results from the exercise in this section on *Daphnia* populations probably can be explained by exploitation and interference elements; the two species involved are both removing supplies and also probably produce mutually harmful toxins. In the companion field exercise on barnacles, one species exploits mutually available space better than the other. Both sets of results should show that with two species in a similar niche, one species eventually 'wins'. This is what Gause (1934) predicted from laboratory experiments on *Paramecium* when he wrote that two similar species hardly ever occupy a similar niche; the niche separation we see now probably reflects the fact that previously competing species have displaced one another, each now occupying a niche in which it performs better than its previous competitor. However, in nature, two or more species do sometimes appear to persist for a long time in an apparently similar niche, although in the laboratory version of their habitat one may exclude the other (e.g. the *Daphnia* exercise). This is probably because conditions in the field are constantly changing so that one species' advantage does not persist long enough to exclude the other.

Competition in plants can involve exploitation competition (e.g. for light, nutrients etc.) but there is also evidence showing that plants can restrict the growth and distribution of other species by producing antibiotic agents. This restriction of the growth of other species is called allelopathy and Exercises 37 and 38 demonstrate this in the laboratory and in the field. Allelopathic interactions have implications for plant spacing and results using techniques in Section Two concerning plants' spatial arrangements could provide circumstantial evidence that allelopathy is occurring.

Predation (and insect parasitism) has received much attention from population ecologists, usually arising out of a desire to know whether predators can regulate the numbers of the prey, i.e. exhibit a proportionate increase in the number of prey killed at higher densities sufficiently 'strongly' to impose an upper limit on the growth of the prey numbers. This interest has largely arisen from the implications of predation for the control of animal and plant pests. That predators usually do not work in this way is partly related to the nature of their response to changes in prey numbers and Exercises 31 and 32 investigate this in detail. Predators do not capture their prey items randomly; their behaviour is influenced by the spacing, activity and physical properties of their prey. Exercises 39–42 should give some idea of how predators respond to prey of different attributes and show some of the selective processes involved in the development of camouflage and mimicry in animals.

Exercises 31 and 32

Responses of predators to changes in the numbers of their prey

Principles

If a predator or insect parasite (parasitoid) is to regulate the numbers of its prey it must cause density-dependent mortality, i.e. it must eat or kill a higher proportion of the prey population at high prey densities than at lower ones. This in itself will not necessarily regulate the prey's numbers but density-dependence is vital for regulation (Solomon, 1976). An increase in mortality by predators could occur (*a*) if an *individual* predator consumed more prey at higher prey densities than at low ones or (*b*) if the *number* of predators increased (through reproduction or immigration) with increasing prey density. The first response to prey density is called the 'functional response' and the second the 'numerical response'. A common type of invertebrate functional response is shown in Fig. 31/32.1 where more prey are eaten at higher densities, but the rate of increase in consumption declines. This type of response cannot alone yield density-dependent mortality because the proportion of the prey

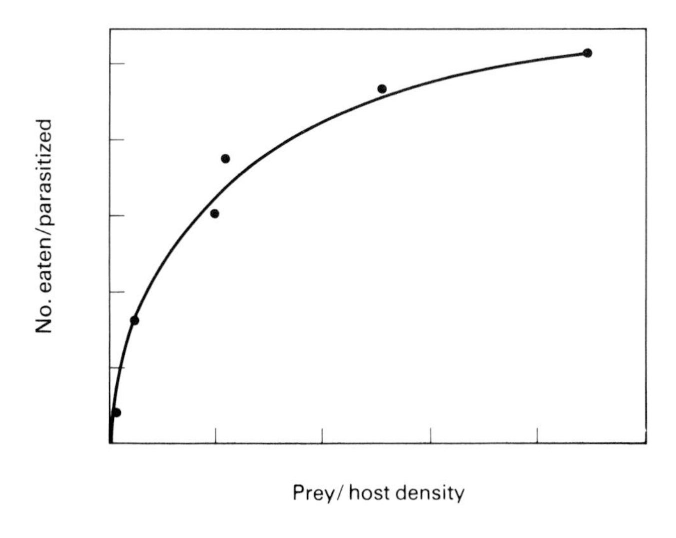

Fig. 31/32.1 A Type 2 functional response of a predator to changes in the density of its prey.

population killed declines with density (Holling, 1959; Murdoch, 1973). A frequent explanation of the type of curve in Fig. 31/32.1 is that at higher prey densities the predator becomes satiated and its proportional consumption declines. However, Holling (1959) demonstrated under laboratory conditions using a human 'predator' that another reason for the declining proportion consumed is that as prey density increases, an increasing proportion of available searching time is spend 'handling' prey items, e.g. capturing, manipulation, feeding, digestive pause etc. As a result, the line does not remain linear but is a curve.

Numerical responses are also varied but when considered alone still need a proportional element before they can form the basis of regulation. They are usually longer term than functional responses and are appropriate for field investigations such as Exercise 32, whereas functional responses are essentially behavioural and can be examined and simulated in the laboratory, as is done in Exercise 31.

Exercise 31
The nature of the functional response of predators to changes in the density of their prey

Apparatus

Eight squares of fibre-board, each measuring 90 cm × 90 cm; 540 sand paper discs of 4 cm diameter or squares of 3.5 cm sides; 8 blindfolds; a clock with a second hand or 8 stop-watches; 540 drawing pins (thumb tacks); graph paper; meter rule; 8 polythene containers (*c.* 200 cm × 100 cm); plasticine.

Preparation

1 Mark the first board with four crosses representing the corners of a square about 20 cm from the board edge.

2 Mark the other boards in regular grids with 9, 16, 25, 49, 81, 100 and 256 crosses; use a meter rule to work out their positions.

3 Pin a sand paper disc, rough side uppermost, in each marked position.

Procedure

1 Working in pairs with one partner as the 'predator' and the other as recorder, carry out the experiment as follows:

2 Select a board at random (i.e. do not start deliberately at the lowest density) and place it on the bench in front of the standing, blindfolded 'predator' (making sure that he has not seen the arrangement of the discs—important at lower densities). Anchor the container which is to receive the discs to the bench to one side of the board with plasticine.

3 For a total searching time of 2 minutes per board (timed by the recorder), the 'predator' *taps* with one finger on the board. When a disc is found, it is picked up, placed in the container and searching is then continued.

4 Carry out a 2-minute 'run' for each disc density, using the same predator throughout.

5 Each functional response (nos. picked up (y) against density (x)) should be plotted, and class data entered on the blackboard. For the class results calculate and plot each mean y and its standard error for each value of x.

6 If Holling's explanation of the departure from linearity is correct, i.e. that an increasing proportion of searching time is occupied in picking up discs as density increases, then the simplest expression of the relationship is:

$$y = aT_s x \qquad (1)$$

where y = No. of discs removed, x = density of discs, T_s = time available for searching. It was assumed that the probability of finding a disc with a tap was proportional to disc density, and a is this constant proportionality factor, i.e.

$$\frac{\text{probability of finding a disc with a tap at disc density } x}{\text{density } x}.$$

If T_t = total time for one experiment and b = time taken to pick up one disc, then

$$T_s = T_t - by \qquad (2)$$

Substituting (2) in (1)

$$y = a(T_t - by)x \qquad (3)$$

which simplifies thus:

$$\begin{aligned} y &= T_t ax - byax \\ y + byax &= T_t ax \\ y(1+abx) &= T_t ax \\ \therefore \quad y &= \frac{T_t ax}{1+abx} \qquad (4) \end{aligned}$$

Equation (4) is 'Holling's disc equation'. For the purposes of using it to fit a curve to the class data, it is advisable to *calculate* the constants a and b. This may be done by transforming (3) into a regression equation of the form $y = \alpha + \beta x$ as below:

$$y = a(T_t - by)x$$

Divide by x:

$$\frac{y}{x} = \frac{a(T_t - by)x}{x}$$

$$\frac{y}{x} = aT_t - aby \qquad (5)$$

Thus when y/x is plotted against y, the slope of the line (β) is $-ab$, and the intercept on the ordinate (α) is $T_t a$.

Therefore, to calculate the constants a and b:

$$\alpha = T_t a$$

$$\therefore \quad a = \frac{\alpha}{T_t} \qquad (6)$$

$$\beta = -ab$$

$$\therefore \quad b = -\frac{\beta}{a} \qquad (7)$$

8 If time allows, substitute the calculated values of α and β and the values of x and $T_t (= 2)$ in equation (*4*) and draw the curve for the class data. If it fits the data reasonably we can assume that the 'handling time' explanation for this functional response is the correct one, as long as a and b really are constant at different densities. This could be shown for b by examining means for pick-up time.

That a must be constant can be shown thus:

The probability of touching a disc at a tap must be equal to the ratio of total disc area to board area.

If: r = disc radius
s = board size
x = disc density

the probability of discovery, P, at each tap will be

$$P = \frac{\pi r^2 s x}{s} = \pi r^2 x \qquad (8)$$

As the number of discs picked up (y) for tapping time T_s must be a product of P, T_s and tapping frequency (i.e. k),

$$y = PkT_s \qquad (9)$$

Substituting (*8*) in (*9*)

$$y = \pi r^2 x k T_s \qquad (10)$$

Comparing this equation with equation (*1*) it is obvious that Holling's factor a is in fact $\pi r^2 k$; if π, r and k can all be assumed to be constant, a is a constant.

Exercise 32
The numerical response of ladybirds (order Coleoptera, family Coccinellidae) to changes in the numbers of their prey

Apparatus

Hand lenses.

Preparation

1 Find one or more areas containing easily-sampled plants with aphids feeding on the leaves and some evidence of ladybird (coccinellid) predation in progress (i.e. eggs, larvae, pupae or adults). Suitable situations would be: 1. a stand of lime trees (*Tilia × vulgaris*) along a roadside, lining a playing field, in parkland or gardens; 2. a similar stand of sycamore trees; 3. several beds of stinging nettles (*Urtica dioica*); 4. wheat or barley in ear with aphids on the leaves or ear; 5. broad or field beans (*Vicia faba*). The best time of year for all these is late spring/early summer.

Procedure

1 The basis of the method is to record the density of predators on a number of plants and to relate this to prey density. Regarding prey and predator numbers as instantaneous measurements a plot of predator numbers (y) on prey density (x) will give a visual representation of the numerical response. By transforming the data to logarithms, one can test whether the numerical response is density-dependent (see below).

2 If sampling trees, select at least ten trees as widely separated as possible, and on each one select 100–200 leaves at random and record the number of aphids, the number of coccinellid eggs, egg-batches, larvae, pupae and adults on each leaf. (Coccinellid eggs are orange-yellow, elongated and laid in batches, stuck to the leaf by one end; a hand lens is necessary to count the eggs in a batch. Larvae are slate grey with paler markings and a sticky disc on the posterior end which they can use to grip the leaf. Pupae are ovoid, dark with paler markings and fastened to the leaf by the tip of the abdomen.)

3 For each tree, calculate the mean number of aphids and the mean number of coccinellid units (i.e. eggs, larvae, pupae and adults combined) per leaf. Plot the mean number of predator units (y) against the mean number of prey (x) so that for ten trees sampled the graph will have ten points.

4 The graph may be curvilinear, which is difficult to analyse, so it is best to transform the data to logarithms and calculate the regression formula and the significance of the departure of the regression coefficient from zero and from 1 (Parker, 1979).

5 If the regression coefficient is significantly greater than zero, one can say that there is a significant relationship between coccinelid numbers

and aphid density; if *b* is significantly greater than 1, then the general numerical response of ladybirds to aphid density is truly density-dependent, i.e. the ratio of predator numbers to prey numbers increases significantly with prey density; if *b* is significantly less than 1, the relationship is inversely density-dependent (Solomon, 1976).

6 If reasonable numbers of several developmental stages of coccinellids are present, the numerical response can be plotted and the regression calculated for each stage in turn. It may then be possible to separate reproductive responses (number of eggs, egg batches, larvae or pupae plotted against prey density) from an aggregative response (number of coccinellid adults plotted against prey density—the adult beetle being the mobile colonizing stage).

7 If herbaceous plants are sampled (e.g. nettles), the principles are the same but instead of using mean number of insects/leaf, use mean number/shoot and calculate the regressions on the basis of a sample of a minimum of twenty widely-spaced shoots.

Discussion and conclusions

It is not unusual for predators and parasites to respond to increasing prey density by eating or killing more prey items, but density-dependent functional and numerical responses are uncommon. By performing the disc experiment (Exercise 31) one can investigate the most usual form of invertebrate functional response and show one reason why it does not yield density-dependent mortality. The exercise shows that:

1 The relationship between prey density and numbers captured by a predator need not be linear.

2 A predator's 'area of discovery' (Nicholson, 1933), i.e. average number of attacks/host, is not constant.

3 A functional response of the above type (Type 2, Murdoch, 1973) cannot yield density-dependent mortality (Solomon, 1976) (shown by plotting *proportion* of prey 'killed' against density).

4 Handling time (which could include manipulation time, feeding time, digestive pause, etc.) can be a major component in the functional response of a predator to prey density, and while prey is being handled, searching ceases.

BUT

5 The model assumes that the hypothetical parasite has no egg limitation.

6 The model assumes that the predator does not become satiated, which is not true in real life. Also, under natural conditions, prey is often aggregated and searching is not always random.

7 The model is not ideal because in practice, as soon as one 'prey' item is removed, the density changes and the true likelihood that the next tap will contact a disc has been reduced accordingly. In other words, if the

total searching time (T_t) is large, the model overestimates the number captured.

8 The disc equation can form the basis of better predator/parasite models if this time element is included, making them true exploitation models (Rogers, 1972). If this is done, however, one equation can no longer predict both predator *and* parasite 'consumption' because of a basic difference in the way they deal with their prey/host, i.e. parasites leave the parasitized host *in situ*, where it could be responsible for wasted time by the same or another parasite re-investigating it.

Further investigations

1 Simple modifications of the laboratory experimental system could be attempted to investigate how marking the discs rather than removing them permanently (i.e. 'parasitizing' them rather than predating them) affects measured handling time and the shape of the curve. Discs could be marked with paper clips and returned to the board; when found again, they should be rejected but left in position (see Rogers, 1972).

2 Arranging the discs in an aggregated way may affect the nature of the curve; this would depend, among other things, on how systematic the searching is. If discs are transferred to the other hand rather than to the box, systematic searching of an area is easier.

3 A simple way of constructing a functional response curve for a real predator involves providing damsel fly larvae (Odonata: Zygoptera) in separate containers with different numbers of *Daphnia*. These larvae are 'lie-in-wait' predators and should each be given a stick to hold on to.

4 The numbers of thrips (Thysanoptera) and predatory anthocorid bugs in gorse (*Ulex europaeus*) should provide data for an aggregative numerical response by a predator to prey density.

Exercises 33 and 34

Interspecific competition in Crustacea

Principles

Interspecific competition in animals is no easier to demonstrate than intraspecific competition (Exercises 25 and 26), especially in the field. Often one sees only the end result of evolutionary competition in the form of one species occupying a 'niche' (see introduction to this section). Alternatively, competition may have occurred in the current generation (e.g. in the establishment of song-bird territories) but still one sees only the end result and not usually the intra- and interspecific confrontations which led to some individuals and species being driven away from the area. A third alternative is that although competition may be occurring between two species, it is so weak that neither is eliminated from the area but the density of each is slightly reduced by the other.

When we suspect that interspecific competition *has* occurred we often have to use circumstantial evidence to support the hypothesis. In the exercise on barnacles (Exercise 34), for instance, there is evidence that both species are physiologically capable of occupying the lower part of the shore but usually only one (*Balanus balanoides*) occurs there. Similarly, when single species of *Daphnia* are cultured without suspected competition (Exercise 33), the populations persist longer than when cultured with the second species.

Exercise 33
Interspecific competition in populations of water-fleas (*Daphnia*)

Apparatus

Sixty litres of artificial pond water, made by dissolving the following quantities of chemicals per litre of distilled water: $KHCO_3$, $NaNO_3$, K_2PO_4, 50 mg each; $MgSO_4$, 24 mg; $CaCl_2$, 80 mg. A few grams of dried yeast; a balance which will weigh 5 mg; one hundred and twenty 150 or 200 ml beakers; paper labels (sticky); rubber bulb pipettes; glass rods for stirring.

Preparation

1 Collect from local ponds sufficient numbers of egg-containing (ovigerous) *Daphnia* of two different species to carry out the replicates

below. These require at least 250 individuals of species A (ideally *Daphia pulex*) and at least 100 individuals of species B (ideally *D. longistigma*). Identify the individuals using the key by Scourfield and Harding (1958). Instead of collecting the required number, each species may be cultured singly by using the concentrations of yeast given below and removing surplus individuals at intervals to new containers with food. In either case, manipulate and identify the individuals using the bulb pipettes, observing the *Daphnia* through the tube.

2 It is possible to complete this exercise in one practical class, but this requires considerable preparation to provide populations at different stages of development at one time. It is easier if students can return to the cultures and sample them at approximately 4-day intervals over a period of 3–4 weeks.

3 Divide the setting up of the following replicates equally among the students. If there are too few students to carry this out in a reasonable time, reduce the number of replicates in each experiment.

Experiment	*No. of replicates*	*Species*	*Water volume (ml)*	*Food level*
1	10	A	100	1
2	10	A	100	2
3	10	A	100	3
4	10	A	50	2
5	10	B	100	2
6	10	A+B	100	2
Total	60			

Food level 1 corresponds to 5 mg of dried yeast per 100 ml of water.
Food level 2 corresponds to 10 mg of dried yeast per 100 ml of water.
Food level 3 corresponds to 20 mg of dried yeast per 100 ml of water.

In experiment 4, 10 mg of yeast are still added even though water volume is reduced by half. Absolute food supply is therefore constant but its concentration is doubled while living space is halved.

Procedure

1 For each replicate, suspend the yeast in the required amount of 'pond' water by stirring vigorously.

2 Use a pipette to add 5 ovigerous *Daphnia* to each beaker. Avoid getting air into the carapace (the 'shell'). Add 5 of each species in experiment 6.

3 Label the beaker with the date, student name, replicate number, water volume, food level and species.

4 Every 3 or 4 days, count the number of *Daphnia* in each beaker. This should be done by removing the animals in batches with the pipette and counting the number in each batch before adding the animals to a fresh culture beaker containing the appropriate water and food quantities. Wash the pipette in 'pond' water before sampling the next culture.

5 Repeat 4 until the population is extinct or has reached an apparently stable level.

6 Plot the development of the populations over time to see how food level, space and interspecific competition affect the growth of the populations.

Exercise 34
Interspecific competition among barnacles

Apparatus

10 cm × 10 cm quadrats (i.e. wood frames enclosing 100 cm^2); string; PVC tape; chalk.

Preparation

1 Find a rocky sea-shore where the barnacles *Balanus balanoides* and *Chthalamus stellatus* occur. In Britain, these species occur together only in the west, from the Isle of Wight round to northern Scotland.

2 Find large boulders or cliffs with nearly vertical faces on which the barnacles occur.

Procedure

1 Allocate each pair of students a rock face on which they should attach the transect-string vertically to extend from well above the uppermost barnacle to below the lowest. Use the tape to attach it to the rocks.

2 Starting from above the highest barnacle, hold the quadrat so that its left or right hand edge is parallel to and touching the stretched string. Mark the position of the lower edge with chalk on the rock.

3 Count and record the number of barnacles of each species in turn which lie wholly within the quadrat.

4 Move the quadrat down the string so that the chalk mark is adjacent to its top edge. Mark the bottom edge with chalk and count the barnacles as before.

5 Repeat until the lowest barnacle has been recorded.

6 If the barnacles extend seaward along the shore below the rock face, it becomes necessary to use a field level and levelling staffs to make sure

that successive quadrats are placed at *vertical* intervals of 10 cm, i.e. a count is made at every 10 cm change in the slope of the shore.

7 Plot the successive counts of each barnacle species as differently-shaded histograms on the same graph to show the zonation of the species with height above sea level.

Discussion and conclusions

The *Daphnia* exercise should support the 'competitive exclusion principle' (see introduction to the section) in that one species should become extinct in experiment 6 while the other persists. Experiments 1–5 should show that the populations attain different equilibrium levels determined by *intra*specific competition for different amounts of food or space. All the populations will probably show an overall slow decline in spite of these trends and this is probably related to the frequent handling and the less-than-ideal culture conditions.

Competition between the two barnacles is not proved by the results from Exercise 34 as they stand, as they could perhaps be explained by a difference in tolerance of desiccation between the two animals. We do know, however, that *Chthalamus* can survive in the *Balanus* zone if the latter is removed but that if both barnacles colonize the middle part of the inter-tidal zone, *Balanus* grows faster and ultimately excludes *Chthalamus* (Connell, 1961a). The upper limit of the *Chthalamus* zone is determined by desiccation and the lower limit of the *Balanus* zone by interspecific competition, not with barnacles but with algae. Predation by molluscs is also a limiting factor here (Connell, 1961b). The exercise does not measure emigration or population trends through time to support the hypothesis that competition occurs, but uses published evidence that the excluded species could survive if the other were not present.

Further investigations

1 Species of aphids cultured as described in Exercise 9 would be convenient organisms for experiments concerning intra- and interspecific competition.

2 Stored-products insects are suitable for competition experiments also but their long generation times require planning well before the day of the exercise.

Exercises 35 and 36

Herbivore grazing as a factor in plant competition

Principles

Grazing by herbivores is an often overlooked factor in the determination of species composition. A classic example of the role of grazers in maintaining a balance was that of rabbits on chalk downland (Thomas, 1960, 1963). Areas where other grazing (e.g. by sheep) was prevented altered drastically when myxamatosis decimated the rabbit population in the 1950's. Likewise, experiments aiming to preserve rare orchid species on chalkland involved fencing areas of downland from sheep with the result that vigorous grasses took over completely, smothering the species the experiments were trying to protect.

When all grazing is excluded from areas one of the major changes to take place is an increase in woody perennials. These species are normally kept in check by grazing animals since they often possess an apical growing point which is repeatedly removed, resulting in stunted growth and eventually the death of the individual. Grasses and many perennial herbaceous species have subterranean growing points or are able to produce new growing points at each internode and are thus less prone to permanent damage.

It is because grazing prevents the natural competitive advantage of woody plants that it has been the standard method of maintaining the characteristic features of commons, heathland and downland through the centuries. It can be considered as a self-perpetuating management regime; the vegetation is managed *by* grazing *for* grazing. However, in certain situations the grazing is not sufficient to prevent the encroachment of shrubs and trees and man has used burning to ensure that woody species do not reduce the areas of effective grazing.

Grazing is thus an ancient method of managing large tracts of land and operates by imposing a halted climax on the vegetation. It is brought about by an alteration in the competitive ability of the plant species in the community. In some areas this type of management is the focal point of controversy since it can be argued that grazing not only maintains the open pasture, heathland, etc., but where grazing animals have access to woodlands they also prevent regeneration of the woodlands themselves. The argument is further complicated by the observation that many 'natural' woodland animals, particularly small mammals, are active grazers and in addition feed on the seeds and fruits of the trees.

The laboratory experiment which examines the effects of grazing on plant morphology takes a number of pot grown plants of several species, both woody and fleshy, and subjects them to simulated grazing over a period of weeks. Since the physical damage treatments to the plants take only a few minutes each week this can be done at the beginning or end of other practicals. During the final laboratory period a whole afternoon is used to compare the 'grazed' with control individuals.

Field situations provide extensive opportunities to investigate the effects of grazing by animals. In Exercise 36 fenced areas are used to exclude grazing animals and a comparison made between the vegetation growing inside these and that outside.

Exercise 35
The effects of simulated grazing on various plant species

Apparatus

Five pots of each species to be examined (see preparation); scissors; wire cutters; dividers; callipers.

Preparation

The objective is to provide plants with a wide range of growth forms (cf. Raunkiaer, 1934) for the practical. Supplies of woody shrubs, e.g. heathers, are readily available at garden centres, as are annual bedding plants. Grasses are easily grown in a greenhouse or in a garden. The final selection of plants and method of propagation depends on individual circumstances, particularly financial, and labour constraints, but native species are to be preferred to garden exotics.

Procedure

1 One month before the practical each pair of students is given five pots, containing the same species of plant.

2 Each pot is labelled with the students' names and numbered 1–5. Each plant is described in detail, e.g. length of aerial internodes, number of leaves, number and position of buds and new shoots, height etc.

3 The plants are treated as follows:

Pot 1. Cut with scissors
Pot 2. Crushed with wire cutters
Pot 3. Broken with fingers
Pot 4. Control
Pot 5. Control.

The severity of the simulated grazing must be adjusted to the robustness of the species used but whatever method is adopted, e.g. top 1 cm, 50% removal, etc., it should be standardized for all treatments.

4 Plants are returned to the greenhouse.

5 On the day of the practical, the students examine their own plants and make comparisons between the treated and control plants. New shoots, position of buds, dead tissue, distance between nodes, etc. can be quickly evaluated with dividers and rulers, and compared with the data collected at the beginning of the experiment. Each student pair should examine in detail the responses of their selected species of plant to the various methods of simulated grazing and comment on their ecological significance.

6 The class results are tabulated as follows

Species	*Growth form*	*Damage*	*Controls*
		1 2 3	4 5

The growth form can be extracted from a flora such as Clapham, Tutin and Warburg (1962) or, if garden varieties are used, from a gardening manual. The life form is then categorized into the standard form proposed by Raunkiaer (1934) which can be found in several ecological texts (Kershaw, 1973; Shimwell, 1971) and numbered in sequence. Damage can be assessed on a visual scale which although subjective and unable to account for reproductive potential can provide a realistic comparison of growth and vigour. The following grazing index scale is one possibility:

0 plant > control
1 no difference between plant and control
2 vigorous new growth but plant < control
3 abundant new growth compensating for > 50% grazed tissue
4 new growth compensating for 25–50% grazed tissue
5 new growth present as aerial shoots
6 very little new growth, e.g. the presence of buds
7 some dead tissue otherwise plant healthy; no apparent new growth
8 plant dead or dying

7 Plot a class graph of damage values against growth form ignoring the differences between methods of simulated grazing. Examine the graph for any overall trends and in particular whether some growth forms are consistently more damaged/retarded than others by ‘grazing’. It should be noted that slow growing species would not be expected to exhibit much response in the time allowed but that does not necessarily mean that such species will not recover—given time. However, for the purposes of this study the objective is to establish whether some plants suffer a greater loss of vigour, and perhaps competitive ability, than others when grazed. Obviously, if a species of plant is slow to recover from grazing (and the vegetation is grazed regularly) then it will be at a disadvantage.

Exercise 36
Herbivore grazing as an ecological factor

Apparatus

100 cm × 10 cm pin frames; measuring tapes; pegs.

Preparation

In an area where grazing by domestic or wild animals is known to be heavy, permanent cages are erected. In woodlands these would need to be the size of fruit cages, whereas in pasture they could cover an area 2 × 2 m. Their construction should ensure that the suspected grazer is kept out. For sheep and ponies a coarse wire mesh would be all that is required and a height of 2 m (preferably with a closed top), whereas for rabbits and other small mammals a finer mesh is required (< 1 cm diameter) and care taken to peg this to the ground, or better still the bottom edge should be buried 20 cm into the ground. Care must be taken with caging since with small areas microclimatic and disturbance effects become appreciable and may influence the results. It is therefore best to build as large a cage as possible.

Where the university or school owns botanical gardens, woodlands or experimental farms then it becomes possible to erect cages as permanent quadrats and study the changes over a period of years. To obtain clear results for this exercise at least one growing season is required and the ideal arrangement is to collect the plant records in the summer term.

Where land is not owned or arrangements cannot be made with owners of large estates it is often possible to study grazing effects by obtaining permission to work on National Park, Forestry Commission or Nature Reserve lands which often have enclosures to restrict grazing.

Procedure

1 Students work in pairs and each pair needs a pin frame.

2 Half the class is allocated to areas where grazing is presumed to take place and half to the enclosed areas or cages.

3 Within the allocated areas each pair of students marks out a 1 × 1 m area within which they work. (If large areas are available, 10 m transects have been found to give better estimates.)

4 Within the marked area the students erect a pin frame and record all the species hit by each point quadrat (pin), for pins 1–10. The frame is then moved to the next position and the vegetation sampled as in Exercise 4 to provide a total of 10 frames per m^2.

5 At half time the students are reallocated so that those examining ungrazed vegetation now examine grazed vegetation and vice versa.

6 Repeat stages 4 and 5.

7 Students should make detailed field observations of the difference

between the grazed and ungrazed areas taking note of, for example, height of vegetation, type of damage to shoots, proportion of grazed shoots, visually dominant species, and degree of flowering.

8 Each pair tabulates their results as follows

Species	*No. of pin hits*	
	Grazed area	*Ungrazed area*
x	n_x	p_x
y	n_y	p_y
z	n_z	p_z
etc.		

9 Collect class results as

Species	*% cover in grazed area*	*% cover in ungrazed area*
x	$\frac{\Sigma n_x}{q} \times \frac{100}{1}$	$\frac{\Sigma p_x}{q} \times \frac{100}{1}$
y	$\frac{\Sigma n_y}{q} \times \frac{100}{1}$	$\frac{\Sigma p_y}{q} \times \frac{100}{1}$
z	$\frac{\Sigma n_z}{q} \times \frac{100}{1}$	$\frac{\Sigma p_z}{q} \times \frac{100}{1}$
etc.		

where q = total number of point quadrats (pins).

10 The differences in composition after one year are usually reflected in the cover values whilst more lengthy experiments result in a change of species composition. To establish the reality of the differences in cover values the Σn and Σp values (not the percentages) are subjected to a χ^2 analysis. Each pair of students is allocated a species and calculates the χ^2 value (see Exercise 3 and Parker, 1979).

Discussion and conclusions

Laboratory experiments simulating grazing effects can become very sophisticated and able to measure subtle differences between grazed and ungrazed individuals. However it is still possible to produce results showing a differential response between various plant species in a short time with simple equipment. In Exercise 35 the differences in the damage are related to life form, which in this case used Raunkaier's method (Raunkaier, 1934) based primarily on the position of the perennating parts of the plant. The results will not tell us which plants will be successful in a grazed environment since only one aspect of the plant is explored. Consideration of the other attributes which may give a plant competitive advantage in this situation will usually involve detailed discussion of growth rates, reproductive mechanisms, both vegetative and sexual, and the ability to withstand successive grazings.

The different types of simulated grazing explore the damaging effects brought about by alternative grazing strategies. Small mammals such as rodents usually have sharp teeth which 'cut' the vegetation since they do not possess the strength to wrench it from the ground. In contrast, cows possess the strength and have evolved a strategy which involves the tearing of vegetation by grasping with the tongue. Sheep can be classified between the two extremes having blunter teeth than the rodents but still able to 'bite' the vegetation producing an effect part way between cutting and wrenching. Other herbivores can be similarly categorized by the nature of their grazing.

Comparisons between grazed and ungrazed plots in the field, as in Exercise 36, allow all the factors influencing competitive interactions between species to operate. Here the results reflect trends which finally result in the exclusion of some species and possibly the survival of new species. In the grazed areas species which can be considered 'weak' competitors because of a slow growth rate or small stature are often able to survive and multiply, whereas larger more robust species may be kept in check or completely eliminated by continual grazing. Examples of these trends can be found in the studies of grazing by Thomas (1960, 1963). He found that tall-growing strongly competitive species such as some shrubs, e.g. *Calluna vulgaris* (heather), and grasses, e.g. *Brachypodium pinnatum* (heath false-brome) and *Festuca* spp. (the fescue grasses), increased in the absence of grazing. Other species fared best with light grazing, e.g. *Helianthemum chamaecistus* (common rockrose) and *Senecio jacobaea* (ragwort).

Using the notes collected in the field in conjunction with the quantitative vegetation records it is possible to make comparisons of the competitive ability of the species examined under two levels of grazing pressure. Although an attempt is made to exclude grazing in the control areas it is virtually impossible to prevent some grazing, especially by invertebrates.

Further investigations

1 Laboratory studies of the effects of grazing are very suitable for long term studies. Methods of monitoring growth rate, respiration rate, etc. of much greater sophistication and precision than used in Exercise 36 can be employed in such investigations.

2 Most plants will be more sensitive to the effects of grazing at some stages of their life cycle/seasonal growth than others. Investigations using a range of species subjected to artificial grazing at monthly intervals during spring and summer would provide a basis for studying the individual species' 'resistance' to grazing.

3 If an experimental garden, field plot, etc. is available it would be possible to conduct experiments parallel to those of Thomas (1960, 1963). Control plots could be left whilst others could be grazed/mowed and the long term changes in both species composition and relative cover monitored.

4 Grazing has a significant effect in determining the vegetational composition of many different habitats. This fact is readily apparent if one considers that grazing operates at many levels, from insects through small rodents to the large ungulates. Most levels will be operating in any given situation. The effect of grazing on the regeneration of trees and other woody perennials is an important feature in restricting their spread. Removal of the growing point during grazing severely retards their growth and successive removal will eventually lead to death or permanent deformation. Studies of regeneration in enclosed and unenclosed woodland (cf. Exercise 20) enable one to enumerate the effects of grazing on woody species and their associated ground flora.

5 Areas which are heavily grazed can suffer nutrient deficiency if the grazing animals are mobile and move in and out of areas, e.g. in and out of woodland, in a daily pattern. Such movements have been observed in deer, ponies and feral goats although little work has been carried out to establish the role of these animals in nutrient recycling of this type. Although the nutrient flow out of a grazed area is not likely to be as intense as a constantly mown lawn where some replenishment of nutrients is essential, it is nevertheless likely to be a measurable phenomenon which may have a significant effect on the vegetation. Experimental plots could be set up to investigate the problem. These might include plots with no grazing; plots with grazing animals penned in permanently; plots where grazers move in and out daily; plots grazed artificially (mowed) with mowings removed and plots subjected to mowing but where the mowings are not removed. Monitoring of the changes in nutrient status and floristic composition of the various plots is likely to reveal the significance of nutrient removal by grazers. Similarly a study of other areas lightly grazed but heavily manured might reveal the effects of a positive nutrient budget.

Exercises 37 and 38

Allelopathy

Principles

Allelopathy is the term used to describe chemical interactions between organisms whereby one organism suppresses the germination, growth or reproduction of another by releasing toxins into the environment. Such interactions can occur between combinations of animals (especially freshwater organisms; Ryther, 1954), plants, bacteria and fungi. The mechanisms of allelopathy are the subject of much current research yet the phenomenon was discovered during the early part of the 19th century. The classic work of De Candolle in the 1830's demonstrated that the failure of continuous one-crop agricultural systems was not in all cases due to nutrient depletion. Studies of 'soil sickness' left unanswered the question of why certain plants failed to grow vigorously when planted on the same site year after year even though the nutrient levels were supplemented with fertilizer and there were no signs of pathogens. De Candolle believed that the cause of sick soil was toxic secretions by the roots of plants. During the early 1900's several experiments confirmed this belief by exposing seedlings to the root washings of suspected allelopathic plants and recording the difference in performance between these seedlings and controls.

Since then numerous studies have demonstrated the widespread occurrence of allelopathy, its importance for agricultural planning and many of the mechanisms by which it operates (Rice, 1974; Garb, 1961; Muller *et al.*, 1968; Muller, 1970; Whittaker and Feeny, 1971).

In the laboratory exercise the investigation examines the inhibiting effect of a crop, in this case barley (*Hordeum vulgare*), on the germination and growth of various weed species. Allelopathy in the field is not as easy to examine, but one of the most common cases is the action of the fairy ring fungus (*Marasmius oreades*) on lawn grasses. It is to be found on many college lawns or in parks.

Exercise 37
The allelopathic action of barley on weed species

Apparatus

In the greenhouse: 50 × 50 cm seed trays to take a minimum of 10 cm soil (John Innes Compost No. 2); seeds of barley and a weed species, e.g.

chickweed (*Stellaria media*). In the laboratory: scissors; paper bags 20 × 10 cm; forceps; balance to weigh 0.01 g; perspex rulers; dividers.

Preparation

For each of the possible variants of this experiment 3 pairs of students are required for one whole afternoon and for a short period one week later. The experimental design is based on the erection of control experiments on the two species (i.e. barley and chickweed) when grown separately and then an experimental growth of the two species in a 1:1 mixture. Throughout the experiment it will be clear that many variations on this basic experiment are possible. For instance, alteration of the proportions of barley and chickweed in the mixture, the substitution of other weed species (e.g. groundsel, *Senecio vulgaris*) for chickweed, staggering the sowing sequence so that either barley or chickweed get a 'head start' or providing a range of sowing densities are all possible with little extra preparation. It is left to the organizer to select the range of variables to be explored, since the management of the exercise is dependent on the student numbers. Therefore the following details describe only the simple situation of 1:1 mixtures of barley and chickweed, but give the necessary information for expansion.

1 During the late summer, collect a large number of seedheads of chickweed. Separate the seed from the chaff and store the seeds in a dry cool place.

2 Two to three weeks before the practical prepare the seed boxes (i.e. for the basic experiment; remembering that alternative mixtures do *not* require further controls unless the species are changed). Fill the seed boxes with the compost to a depth of 10 cm and level and firm the surface.

3 Mark on the surface of the soil a 10 × 10 grid (i.e. 5 × 5 cm squares) and in the centre of each square firmly place alternately a barley and a chickweed seed. This arrangement may also be varied to provide alternate rows of barley and chickweed, random mixing or blocks of each species.

4 Cover seeds with a light sprinkling of soil, and water with a very fine rose or mister. Mark each tray with details of the experimental design and students' names if they have carried out this preparation.

5 The experimental plants are kept at around 20°C (and preferably 12 hour daylight conditions) for 2–4 weeks.

6 To act as a further control, barley and chickweed seeds are grown separately at the same density (i.e. 1 seed per 5 × 5 cm square) and the chickweed watered with water that has passed through the barley. This can be achieved by collecting the run off in a photographic developing dish (larger than the seed tray) placed under the seed tray and using this to water the chickweed. Barley can also be grown in water culture and this water, changed daily, used to water the chickweed. This control excludes the possibility of physical interference.

Procedure

1 Each pair of students is allocated a tray.

2 The students record a number of features of the plant(s) in the tray. The selection will vary depending on weed species but the following have been found suitable

Percentage germination
Mean leaf length
Mean internode distance (between set nodes) both for stem and root
Mean height
Mean number of flowers per plant

3 The mean dry weight is determined as mean dry weight of aerial shoots since it is difficult to separate the fine roots. Ten plants (of each species if it is a mixed tray) are cut off at ground level, washed, blotted dry and placed in the paper bags which are pre-weighed and marked in pencil. The plants may be cut into small pieces to fit the paper bag as only the total weight of 10 plants is required.

4 The paper bag is placed in an oven at 60°C for 1 week.

5 The paper bag is removed from the oven, allowed to cool and weighed.

6 The mean dry weight of plants is calculated from

$$\frac{\text{wt. of plants} + \text{bag} - \text{wt. of bag}}{10}$$

7 Compare the morphological and weight characteristics of the chickweed when grown separately, when grown with barley and when watered with barley washings.

8 It is possible to compare values statistically by employing a 2×2 contingency table (p. 14) for those cases where data can be categorized, e.g. germination/no germination. For continuously variable data, e.g. leaf length, a *t*-test can be used to test the significance of the comparison (see Parker, 1979).

Exercise 38
Field investigations of allelopathic interactions

Apparatus

25 m measuring tape; scissors; dividers; self sealing polythene bags 20 × 10 cm; self adhesive labels 2 × 4 cm.

Preparation

1 During the autumn quite distinct rings (fairy rings), often broken but sometimes complete, of the fungus *Marasmius oreades* are found on

ornamental lawns, reclaimed land, playing fields and waste ground. The position of several good examples should be recorded on a small scale map. Samples of the fungus fruiting bodies are collected for detailed identification since other basidiomycete fungi can cause similar rings (Ramsbottom, 1953).

2 One week before the practical session the site is revisited and the fairy rings located. This is usually quite simple, for although fungal fruiting bodies may be absent the grass is much darker around the circumference of the ring.

Procedure

1 Students work in pairs and are arranged to enable them to lay a transect, bisecting a fairy ring. For small rings 1–2 m across only one pair of students should be allocated per ring to avoid trampling and other physical disturbance. For the largest rings > 10 m diameter several pairs can work on the same one.

2 Transects are constructed using the 25 m tapes. Students should aim to position these transects across the diameter of the ring. Obviously in some cases it will be difficult or impossible to detect a regular shape or complete ring. In such cases it is important to designate part of the transect as 'within the ring', and part as 'the ring' (i.e. the dark grass area) and part as 'outside the ring'.

3 A species of grass is selected for close examination and at 10 sampling sites (the length of transect/10) morphological characteristics such as the average height of 10 individual plants, the mean number of spikelets (at the right time of year), the length of leaves or length of internodes are recorded (see Exercise 37).

4 At the 10 sampling sites an area of 10 × 10 cm is clipped with scissors to ground level and all the aerial plant material placed in labelled polythene bags.

5 On return to the laboratory the clipped material is washed, blotted and then dried in preweighed paper bags. The bags can be left in an oven at 60°C for 1 week. If they are stacked, either spacers must be used to ensure good air flow or the bags randomized twice every day.

6 Weigh the dried plant material in the bags and calculate the biomass for each sample site by subtracting the weight of the bags from these final weights.

7 Plot the biomass and morphology results against distance along the transect. Mark the distance axis also with the position of the advancing front of the fairy ring and designate the areas either side as inside or outside the ring.

8 If all the students have recorded data for the same species, then group the class biomass data into two blocks

(A) Biomass records collected inside a ring

(B) Biomass records collected outside a ring

To test whether there is any overall significance between the performance of the lawn species inside and outside the ring a *t*-test is employed.

9 If several species have been examined, compare and contrast any trends in morphology or biomass with reference to the growing front of the fungal ring. Are some species more affected by the fungus than others? What effect does the size of ring have on the results?

Discussion and conclusions

Barley, like other 'smother crops' such as rye, sweet clover, millet, soya, sunflower and alfalfa, has been known for many years as a suppressor of weeds. Overland (1966) conducted the first controlled experiments investigating the interaction between barley and chickweed and suggests that there is a root exudate of the barley causing germination—and growth—reduction in chickweed. The precise chemical action is not understood but the active inhibitory agent, the allelochemical, was found to be an alkaloid.

Allelopathy is found in many field situations but is usually more difficult to analyse than in the case of the fairy ring fungus *Marasmius oreades*. However, in tropical and desert habitats examples are widespread. In the latter the need for reducing competition for water is obvious and some plants impose regular spatial pattern on their own community by releasing toxins which inhibit the germination of both their own and other species' seeds. Went (1942) demonstrated this inhibition with the desert shrub *Encelia farinosa* which suppresses desert annuals. Likewise Muller (1966) studied the aromatic shrubs of the Californian chaparral and found that volatile terpenes were released from the leaves of plants such as *Salvia uncophylla* and *Artemesia californica*. These chemicals appear to inhibit the growth of grasses in the field which results in zones 1–2 m wide of bare ground around each shrub. Hence it would be possible to conduct numerous parallel exercises to the *Marasmius* one above in a wide range of habitats.

The action of the fungus *Marasmius oreades* is to grow outwards from a central germination site, the hyphae moving outwards in all directions and producing a ring of fruiting bodies at the extremities each year. As the mycelium grows and metabolizes it releases several toxic substances into the soil, including cyanide and other antibiotic substances. These are thought to act on plants directly and on soil microorganisms, having the secondary effect of inhibiting nutrient recycling. However, at the advancing front either enhanced nutrient levels or growth promoting substances produce a high growth rate in the grass and other lawn species.

Further investigations

1 It is possible to conduct the investigations on the allelopathic effect of barley seedlings on weed species in an experimental garden. If that is

possible then it is best to leave the plants for 1 month, or even longer if the weather is cool. The results obtained when the plants are left longer are easier to interpret since a higher proportion of the weed controls should germinate and for many of the 'opportunist' weed species the number of flowers can be used as another parameter. Barley significantly reduces flowering in some weeds. This phenomenon could be explored as a long term project.

2 Development of the greenhouse version of the barley experiment would enable variables to be controlled and manipulated. For instance the roots of each plant could be separated by using a gridded seed box such that the roots were partitioned off; both root and shoot could be partitioned or alternatively the shoots alone could be separated.

3 A parallel line of investigation to 2, above, would be to produce extracts of roots and shoots of barley to identify the parts of barley plants which contain the growth inhibiting factor. Similarly root washings or even water passed through barley cultures could be tested on various weed seedlings.

4 The field investigation of allelopathy offers a wide range of possibilities in many habitats. Orchards have been the subject of several studies of allelopathy (e.g. Proebsting, 1950). These studies concentrate on the effect of fruit trees on subsequent plantings of the same species. Fruit trees are also affected by other species, for instance apple trees are inhibited by the presence of several species of grass (Pickering, 1917). It should be possible to formulate both laboratory and field investigations to explore the effects of lawn turf on a wide range of agriculturally important plants. The work of Pickering involved passing water through turf which was suspended above the seedlings and erecting controls which consisted of (*a*) water applied to the seedlings which was passed through soil only and (*b*) turf held above seedling cultures but receiving independent water supplies.

5 A wide range of physiological and biochemical experiments is possible in the study of the allelochemical nature of fungi. These would require careful laboratory culturing of the fungus, which is often difficult, and the control of the many variables influencing the soil ecosystem. Studies in the field would be more limited but offer the possibility of studying the effects of allelopathy on lawn species. Physiological changes in the plant tissues of individuals of lawn species influenced by the exudates of *Marasmius* could be compared with 'normal' individuals. Cytological techniques of examining the fine structure, or biochemical techniques which may expose differences in nitrogen or chlorophyll content between the two sets of individuals, offer a wide scope for project work.

6 Many fungi have been observed as having a growth suppressing effect on neighbouring organisms. As already mentioned, several Basidiomycetes can affect vascular plants on lawns, and on large commons and ornamental grasslands it is possible to compare the toxic effects of different species.

7 In laboratory studies various fungi can be cultured and their effects on soil bacteria and other fungi investigated (see Jackson and Raw, 1966). In woodlands where fallen timber is left, a wide variety of fungi are found. The patterns of colonization of logs can be studied on sections of trunk with reference to the presence or absence of certain species of fungi. This type of investigation requires careful preparation to identify species or that the investigation be performed in the autumn when the fruiting bodies are available both to locate and identify the fungi.

8 Bracken (*Pteridium aquilinum*) exhibits strong allelopathic effects on other species and could form the basis of several field or laboratory projects (see Perring, 1974).

Exercises 39 and 40

Batesian mimicry

Principles

Mimicry is the resemblance of one animal (the mimic) to another (the model) so that a third animal (usually a predator) is deceived by the resemblance and confuses the two. The theory of Batesian mimicry states that the model is unpalatable and therefore avoided by predators and that the mimic, although edible, gains some protection because it resembles the model; it is preyed upon less than if it was not a mimic. For instance, hoverflies (Syrphidae) are Batesian mimics because they mimic the black and yellow banding of wasps.

The implications of Batesian mimicry are:

1. The mimic must not become too common relative to the model; if it did, the predators would develop a preference (see Exercise 41) for the pattern rather than a conditioned avoidance.

2. If some predators can overcome the noxious characteristics of the model, then the mimic will also be predated.

3. Some predators may detect the differences between a less-than-perfect mimic and its model, and then use the mimic's bright colours to find it more easily than if it were non-mimetic. This will occur especially when the mimic is relatively common (see 1 above).

4. For Batesian mimicry to evolve gradually, by the accumulation of a succession of favourable mutations, there must be some selective advantage in being even a poor mimic. In other words, predators must be able to generalize from a poor mimic to the model and predate the mimic less than a non-mimetic form, but more than if it was a good mimic.

In the laboratory exercise we construct an artificial community of models, mimics and a non-mimic to investigate whether 'predation' of the mimic (in this case using a human 'predator', as in Exercise 31) is frequency-dependent. In other words, is the mimic relatively better off when rare in comparison with the model (items 1 and 3 above)? We look for evidence that the proportion of the mimics eaten, relative to their frequency in the population, is greater when the mimic comprises a large proportion of the community than when it is rare.

Exercise 39
Is Batesian mimicry frequency-dependent?

Apparatus

Four hundred squares or circles of card (the thicker the better) about 3 cm in diameter; graph paper; as many packs of playing cards as possible; one tally-counter per group of 3 students; a clock with a second hand.

Procedure

1 Students should work in groups of three. In each group one student should be kept unaware of the purpose of the exercise and should not observe the marking and distribution of the paper squares (see below).

2 The paper squares should be marked with a chosen symbol to simulate 5 prey species, A to E. B mimics A and 100 of each should be marked with a circle; B is 'palatable' and should have an additional mark (a *lightly*-pencilled cross on the *underside*) to signify this. If the pencil is pressed too hard the mark will be visible from above. D mimics C and 45 of each should be marked with a new symbol, the mimic also receiving a cross for palatability on the underside, like B. E represents a different non-mimicking, palatable species and should be marked with a cross above *and* below.

3 Each student group should carry out 13 trials in which the proportion of the model A and mimic B are varied but those of C, D and E are kept constant to serve as ecological 'background noise'. These should ensure that the 'predator' remains unaware of the purpose of the exercises. In each of the 13 trials, 45 C cards and 45 D cards should be scattered on the bench with 10 E's. The numbers of A and B in each trial should be as follows, randomized on the bench top with the C, D and E prey items:

Trial	1	2	3	4	5	6	7	8	9	10	11	12	13
A (*model*)	100	95	90	80	70	60	50	40	30	20	10	5	0
B (*mimic*)	0	5	10	20	30	40	50	60	70	80	90	95	100

4 Each group should carry out the trials in random order. Separate a complete suit from a deck of cards, shuffle them and lay them face up on the bench; their sequence gives the order in which the trials are carried out.

5 Recall the predator and tell him/her:

(*a*) that some of the prey are palatable (have crosses on the underside) and some are not (no crosses on the underside).

(*b*) that he/she must 'eat' as many palatable prey as possible in the time allowed ($1\frac{1}{2}$ minutes). To 'eat' a prey item, the 'predator' should click the tally counter and hand the prey to the recorder. Un-palatable prey should also be handed to the recorder. A prize for the predator who catches most palatable prey in a given trial might increase motivation!

(*c*) that he/she is only allowed to deal with one prey item at a time.

(*d*) that he/she must not concentrate on one small area.

6 Start the clock and allow the predator to 'feed'. As the prey items are handed to the recorder they should be lined up in sequence on a different part of the bench and the prey sequence recorded at the end of the $1\frac{1}{2}$ minutes (in the absence of the 'predator'). Record also the proportions of each of the 5 species attacked.

7 Adjust the proportions in the community for the second trial, recall the predator and repeat the procedure. As pieces of card become bent or damaged, replace them with similarly-marked pieces.

8 Repeat 6 until all 13 trials are completed or until the time available is used up.

9 The 'predator' can now participate in the analysis. Plot the data for all groups on one graph as follows

$$y \text{ axis:} \quad \frac{\text{No. of B eaten}}{\text{No. of (A+B+C+D+E attacked)}} \times 100\%$$

$$x \text{ axis:} \quad \frac{\text{No. of B offered}}{\text{No. of (A+B+C+D+E offered)}} \times 100\%$$

10 Draw the line of equality ($x = 30\%$, $y = 30\%$ etc.) and estimate whether this is a good description of the results. Look for evidence of frequency-dependence. In the data collected on the sequence of predation, look for evidence for changes in the number of C eaten and for evidence that the prey were taken in 'runs'.

11 It is quite likely that for the low values of x, the results may fall mainly *below* the line while at high proportions of the mimic in the community they may fall *above* the line. However, to test this we cannot calculate a regression of y on x and compare the slope with $b = 1$ because, firstly, the data are proportions. Secondly, if the relationship was sigmoid, the regression might 'average out' the curve and give a slope of one. However, we can use a non-parametric method (which does not depend on frequency distributions of the data) such as the runs test (see Exercises 17 and 18). For each frequency of B offered, calculate the class

means for the proportion predated. Classify these means as 1 if they fall above the line on the graph and 0 if they fall below it. The above method tests the significance of the divergence of the 0–1 sequence from a random order.

Exercise 40
Do slight resemblances to unpalatable models have survival value for a Batesian mimic?

Apparatus

For every 6 students: 80 card triangles prepared as below. Colours should be checked against freshly-made cards as some fading in sunlight may occur. For the whole class: 2 kg flour/lard pastry (uncoloured) made into baits as in Exercise 42. A saturated solution of quinine dihydrochloride (*c.* 5% at room temperature), obtainable from most chemists; waterproof drawing ink of the following colours: black, red, yellow and green; small paint brushes; 40 stakes or pegs *c.* 10 cm long; a tape measure at least 5 m long.

Preparation

1 Mark out with the stakes an area of mown grass 5 m × 5 m where ground-feeding birds are known to occur. Divide each side of the square so formed into 10 half-metre sections with the rest of the stakes. Prepare such an area for every 6 students expected to take part in the exercise.

2 Three weeks before the exercise, provide food (bread, scraps) in the squares at the same time every day for a week to condition the birds to visit the areas.

3 For each square, prepare 80 equilateral triangles from stiff card, each with a height of 3.5 cm. Mark each card with a 0.5 cm black border. This may help to define the cards against the background. Colour the cards as below

1. 32 green. Controls—to receive edible pastry.
2. 16 red. Models—to receive pastry soaked in quinine dihydrochloride.
3. 8 red, with a black dot on the *underside*. Perfect mimics; pastry with no quinine.
4. 8 red, with a 0.5 cm horizontal black bar. Imperfect mimics; pastry with no quinine.
5. 8 yellow. Imperfect mimics; pastry with no quinine.
6. 8 yellow with a black bar. Very poor mimics; pastry with no quinine.

4 Two weeks before the exercise, distribute 32 green cards, each with a pastry bait placed centrally, in each of the marked squares. Distribute the cards more or less randomly in the squares. These baits are controls and are to condition the birds into accepting a novel food source. Record the

time taken for all the baits to be eaten. This will probably decline over the first few trials, then level off. Replace the baits at least daily, preferably more frequently.

5 Eight days before the exercise, for each square soak *c.* 300 baits in quinine solution for 1 hour and let them dry.

6 The next day, distribute 16 controls and 16 models randomly in the square. Designate the half-metre divisions on the square's sides 0–9 and use random number tables to give coordinates for placing the 32 'prey' items. The models are the 16 red cards (without a black dot below or a bar above) on each of which is placed a quinine-soaked bait. When the time representing a stable rate of removal in 4 above has elapsed, record the number of each type eaten. Repeat the experiment daily until the day of the exercise.

Procedure

1 Each group of students should set up 32 baits randomly in their square. Use 8 controls (green), 16 models (red with quinine baits) and 8 of *one* of the 4 mimics, chosen randomly.

2 Leave the experiment for about the standard time determined by the pre-experimental conditioning, recording during this time the birds' reactions to the baits. The time can be modified if fewer birds are present.

3 Record the number of each bait eaten. Classify those pecked but not consumed as uneaten, but record their number.

4 Repeat the experiment with a different mimic, recording as before.

5 If the birds' feeding rate is so slow that not all mimics can be tested in the time available, the random order of presentation of the mimics by each group should ensure that all mimics have been presented with more or less similar frequency by the class as a whole.

6 Tabulate the numbers of each bait eaten for all the class experiments and carry out analyses of variance for each control/model/mimic combination to look for evidence of lower rates of predation on the model and perfect and imperfect mimics. An analysis of variance will not always be strictly appropriate for these data, but the reasons why non-parametric tests should sometimes be substituted are too complex to be dealt with here.

Discussion and conclusions

The laboratory exercise should provide good evidence for frequency-dependent selection of a mimic. If this is true of real populations, we can conclude that a mimic is at a selective advantage when rare relative to the model. When the palatable individuals are outnumbered by a similar but distasteful model, the predator becomes conditioned to avoid animals of that general appearance because most feeding attempts result in failure and an unpleasant stimulus. As the mimic becomes more common, the predator is more likely to search actively for the common pattern because the chances of encountering a palatable mimic increase. This type of frequency-dependent selection has been shown to be true for real predators and prey (starlings (*Sturnus vulgaris*) and mealworms (*Tenebrio molitor*)) by Brower (1960), although these experiments were only semi-natural in that they were performed on caged birds. They did show, however, that there was a detectable advantage to the mimic even when its frequency in the total population was high (90%).

For Batesian mimicry to evolve we would expect poor mimics to be at some advantage compared with non-mimics. The results from the field exercise should support this and show that predators (in this case ground-feeding birds) can generalize. This exercise can provide a great deal of extra information if recording of the birds' behaviour and predation rates is done in an exploitive way.

Further investigations

1 The relationship in Exercise 39 would probably differ if the palatable mimic was made less than perfect (e.g. the circle marking prey B could be placed slightly off-centre).

2 Natural prey items could be substituted for the pastry in Exercise 40. Mealworms killed by low temperature and then dyed and soaked in quinine could be used.

3 An animal which mimics poorly a model which gives a predator an only mildly unpleasant experience when it is captured (e.g. a mildly unpalatable prey) might be expected to receive less protection than a similar mimic of a very unpalatable model. This could be tested using different concentrations of quinine dihydrochloride in Exercise 40.

4 The chapter on Batesian mimicry in Edmunds (1974) includes many ideas which could be tested or extended using modifications of the above two Exercises.

Exercises 41 and 42

Camouflage and apostatic selection

Principles

Many species of animals and plants exhibit genetic polymorphisms in which two or more varieties differ structurally and/or in their colour markings. These 'morphs' may also occur in different proportions in natural populations. It has been suggested that when a predator which hunts by sight preys on a species which has a common and a rare morph, the predator will concentrate its search on the common one and tend to overlook the rare one (Clarke, 1962). This would mean that the selective value of a rare morph is inversely related to its frequency and predation acting in this way would favour its persistence. Clarke described such selection as 'apostatic' because it favours apostates, i.e. those individuals which are visually distinct from the norm. This hypothesis can only be valid if predators do concentrate their attention on common prey varieties.

One way we can examine a population of wild predators for evidence of apostatic selection of prey is by first training the animals to take an abundant prey item and then presenting this to them in equal proportion to a prey they have not seen before. Exercise 41 does this, following the technique of Allen (1974). If a significantly higher proportion of the prey with which the predators are familiar is taken, we can conclude that the predators hunt in a manner that can maintain polymorphism. The experiment is analogous to a situation where one prey is very common. If this is so, then by chance the predator is likely to encounter a number of them before a rare one and thereby form a preference for the common type.

It could be argued that the experiment with pastry baits in Exercise 42 is artificial but the results from the laboratory exercise should show that predators (in this case thrushes) do consume polymorphic prey in proportions different from those occurring naturally in the prey population. This was shown clearly by Sheppard (1951), although in the latter case selection was *against* the apostate.

Exercise 41
Prey camouflage and apostatic selection in wild birds

Apparatus

A collection of the snail *Cepaea nemoralis* and shell fragments from 'anvils' of song thrushes (*Turdus philomelos*)—see below.

Preparation

1 Using local knowledge and personal searching, locate in early spring a population of the snail *Cepaea nemoralis*. This species favours alkaline soils and commonly occurs in beechwoods in the south of England. Other polymorphic species such as *C. hortensis* could be used. It is also necessary to locate as many thrush anvils as possible.

2 Collect the snail shell fragments from each anvil at 3–4 day intervals. Label them with the location from which they were collected. Make collections from early spring onwards.

3 Every two weeks from the first emergence of the snails, search for as many living ones as time allows in the vicinity of each anvil. The morphs vary from individuals which are unbanded to types possessing 1–5 bands. Record the proportion in each morph of the living snail population in the vicinity of each anvil site. Continue the counts and collections of snail fragments into early summer.

Procedure

1 Provide each student pair with a dated collection of shell fragments together with a note of its origin.

2 Each student pair should try to decide how many individual snails are represented in the sample, the morph proportions and which, if any, are without banding. Confirm the snail species by referring to the simple key in Cloudsley-Thompson and Sankey (1961). In this book, *Cepaea* is called *Helix*. Alternatively, use Kerney and Cameron (1979).

3 Enter on the blackboard the number of snails in each morph in the sample, together with its date.

4 The proportion of snails in each morph should be calculated for the shell fragments from each anvil site for each collection date.

5 The proportion of shells in each morph at the anvils should be plotted against time, together with the data on morph frequencies in the living population, the latter provided by the organizer.

6 Look for trends in the proportions in the morphs on anvils and in the live population. Are they similar?

Exercise 42
Apostatic selection of artificial prey by wild birds

Apparatus

36 metal or wooden pegs; pastry baits prepared as below; random number tables; a pack of playing cards with the kings removed; a sheet of perspex 10 cm × 10 cm with *c.* 10 holes, each of *c.* 0.7 cm diameter drilled in it (or a child's 'Play-doh' machine).

Preparation

1 Make the pastry baits as follows: knead 1000 g plain flour with 400 g lard to give a dough of plasticene-like consistency. Mix 10 ml brown food dye into the dough until it is uniformly coloured.

2 Force the dough by hand through the Perspex sheet to produce 'sausages' of pastry. Lay these side by side and chop them into sections approximately 0.7 cm long. Store the baits in plastic boxes.

3 Repeat 1 and 2 but use green food dye and double the quantities of the ingredients. Make more dough if necessary during the training or experimental periods.

4 Mark out a 5 × 5 grid of 1 metre squares on a mown lawn or on bare soil using the pegs to mark the corners of each square. Choose an area where blackbirds or other ground-feeding birds are known to feed.

5 6–7 days before the exercise is due to begin (at anytime of the year other than the period when passerine birds are feeding young), arrange green and brown baits as in 2 below and record and replace eaten baits as described for student participation in 3 below. This stage is to find out if the birds already have a preference for a particular colour (see Bantock and Harvey, 1974).

6 At the end of the 24 hours, remove the remaining baits and scatter 300–400 *green* baits over the grid. Roughly top up the number of baits daily, up to and including the day before the exercise.

7 Draw a plan of the experimental plot, with each square divided into 4 quadrants.

Procedure

1 On the first day of the exercise a pair of students should remove any baits remaining in the experimental plot.

2 Distribute 25 green and 25 brown baits in the plot so that each square receives two baits. For each square in turn, one of the student pair should mentally divide the square into 4 quadrants and cut the deck of playing cards twice to decide which colour baits should be placed in the square and in which quadrant. A cut producing cards with a face value of 1, 2 or 3 signifies the first quadrant, 10, Jack or Queen the fourth etc. Let a black suit signify a brown bait and a red suit a green one. The

second student places the baits in the appropriate places and records a G or B on the plan of the plot. If the same quadrant by chance is selected twice in the same square, cut the cards again.

3 For the next 24 hours, during daylight, replace eaten baits frequently with reference to the plot plan (i.e. at intervals of *c.* $\frac{1}{2}$ hour, depending on the feeding rate of the birds). A third, new, student can do this. Record the number of baits of each colour which is replaced at each visit, together with the time of the visit. Half way through the daylight hours before the next experiment, a fourth student should take over the recording and replacing.

4 Twenty four hours after the experiment was set up a second student pair should take over the recording, starting with 50 baits placed according to the original arrangement.

5 Repeat 3 and 4 for up to one week then plot the proportion of brown:green baits taken for each visit against time.

6 Look for a trend in the relative numbers of baits eaten through time. Carry out a χ^2 analysis (Parker, 1979) on the numbers taken at the first visit by a recorder compared with the last, or make other suitable comparisons.

7 Do the relative proportions of green and brown bait eaten differ between the trial run and those eaten on each day of the student exercise? Use χ^2 to analyse this.

Discussion and conclusions

Exercise 42 should show that the birds had developed a preference for the familiar colour and that the tendency to search for this colour decreased with time. The latter result appears to indicate that the preference wanes with time and needs reinforcing. If the suggestion below is followed, where the birds are re-trained with the alternative colour, it should also be possible to show that preferences are reversible. In every case some birds will probably have consumed some of the unfamiliarly coloured 'prey', even on the first day following the training period. This is probably related to immigration by new birds which were not present during the training period and to the likelihood that some birds will have found, by chance, a series of brown prey items and developed a preference for brown, despite the previous green training. Apostatic selection by predation of the familiar morph has been demonstrated for other wild passerine birds only with artificial food (e.g. by Croze, 1970) and not for predators trained to search for a particular morph of a natural polymorphic species. Although the results from the laboratory exercise should show a difference in the proportion of yellow snails on anvils and those in natural populations, this is not apostatic selection because it is acting to *diminish* the amount of variation. The birds' choice of prey is probably mainly related to its visibility against a background which changes as the season progresses; in early spring, yellow snails would be expected to be conspicuous against the soil which has not yet developed a cover of herbaceous plants. In late spring, yellow-shelled snails which appear greenish-yellow with the living animal inside, are probably inconspicuous against the background of new vegetation.

Further investigations

1 If time allows, after *c.* 3 days have elapsed during which the birds in Exercise 42 have been exposed to a population of green and brown baits, begin a new training period using brown as the training colour. Follow this by presenting the bird population with a 1:1 mixture as before. This should demonstrate that preferences are reversible.

2 Apostatic selection involving wild birds can be investigated in a different way from that in Exercise 42. Allen and Clarke (1968) exposed *untrained* birds to populations of green and brown baits in the ratio 9:1 or 1:9 and found that although brown was preferred in both populations, the preference was stronger when brown was the common colour. This experiment could easily be repeated and extended.

3 The results of Exercise 42 apply to dispersed populations but in very dense populations one might expect the rare morph to be eaten with a greater than expected frequency. If this happened, it could be explained by the predators selecting individuals which are conspicuous with respect to their neighbours. Such predation would not therefore lead to apostatic selection favouring the polymorphism, but would promote uniformity of appearance. Allen (1972) demonstrated this effect of 'prey' density and, like the dispersed bait experiments, this investigation could be repeated and extended as a field exercise.

The populations should be spaced more widely than in Allen's work however, e.g. 1 km apart, otherwise complications arise concerned with the birds moving between populations.

Section Five

Community Analysis

Introduction

If we consider a community as a set of populations of plants and animals in a defined place possessing some degree of integration, it follows that analysis of a community should concentrate on the identification of the populations concerned and their degree of integration. In practice the plant ecologist has concentrated on the detection of naturally occurring groups with much emphasis on the analysis of species composition. The description of vegetation is based almost entirely on methods of community analysis and many of the current and historically important techniques are described in Shimwell (1971). Here we will investigate only a selection of techniques requiring numerical analysis.

Another property of communities often used as a tool for comparison is the relative commonness or rarity of their constituent species. Exercises designed to provide experience in analysing populations for these components are found in this section, as are exercises which explore the concept of diversity in both animal and plant communities.

Integration between the populations of a community is difficult to demonstrate in a short practical, but an attempt is made in the following exercises to demonstrate how groups of animals and plants can be analysed to identify and enumerate relationships between them. Other exercises establish associations between species, which does not of course necessarily mean that a dynamic relationship exists between the organisms concerned, since they could all be related to some habitat factor which causes them to occur together, e.g. parasites of sheep and calcicole plants are dependent on the habitat rather than each other.

Communities change with time and species succeed each other in a natural progression. This is true of both plant and animal communities, as can be seen for example in hydrosere development in plants and the colonization of carrion in animals. Thus, certain communities can be characteristic of a stage in a seral progression. Similarly, palaeoecology can provide a background to understanding present community relationships. Because of the length of time required to collect the detailed data required for community analysis or the numerical processes involved, several of the exercises cannot be completed in a single afternoon. However, they are suitable for two sequential afternoon sessions, one used for data collection and one for analysis and interpretation.

Exercises 43 and 44

Diversity

Principles

The concept of diversity is usually divided into plant diversity and animal diversity although an increasing number of biologists (especially those investigating whole ecosystems) are attempting to quantify community diversity. This plant/animal division of diversity is often arbitrary, usually reflecting the background and interests of individual ecologists. However, there are some intrinsic differences between plant and animal components. For instance, plant diversity is often stable throughout the year whereas animal diversity tends to fluctuate over a much wider range. This is mainly because plant diversity measures do not usually suffer the rapid fluctuations due to migrations of organisms which can occur in animal communities.

Species diversity at its simplest level is a comparison of species richness, i.e. the number of different species present in different communities. Most ecologists are not satisfied with this definition since it takes no account of the number of individuals of each species present. Two plant communities, for example, could each possess twenty different species of flowering plants, but in one community there might be ten individuals of each species, whereas in the other there might be several hundred of one species but only one specimen of each of the others. It thus becomes clear that it is more desirable for any measure of diversity to take account of both species number and species abundance. There are several indices that express both these properties and in the following exercises one of the standard techniques, the Shannon-Wiener Index (McIntosh, 1967), and a rapid technique known as the Sequential Comparison Index (Cairns *et al.*, 1968) are used.

Exercise 43
Diversity of freshwater invertebrates

Apparatus

Binocular microscope; eye dropper/pipette of *c.* 5 mm diameter; freshwater nets, e.g. those marketed by the Freshwater Biological Association; plastic buckets (with lids) *c.* 5 l capacity; white plastic photographic developing trays; fine paint brushes.

Preparation

1 Collect freshwater organisms from a number of different habitats by employing a system of standardized sweeps. The exercise has been used to study changes in diversity both along single river systems, i.e. by sampling in the head streams, in quiet backwaters, in rapids and in smooth glides, etc., and between different freshwater habitats, e.g. streams, lakes both exposed and sheltered, farm ditches, acid and alkaline rivers, etc. In most areas it is possible to collect from 5–10 different habitats or, as an alternative, the exercise can be used to compare two freshwater habitats in greater detail. In this case, sampling is carried out at ten different stations in each of the two habitats.

2 Place the catch from each station/habitat in a plastic bucket containing water from the same source, and, if possible, a quantity of aquatic vegetation.

3 Label bucket.

4 If the organisms are not to be used immediately then they must be stored in a cool place and, if from running water, the water in which they are kept must be oxygenated.

5 Where fast-moving or predatory invertebrates are likely to cause problems, the organisms may be immobilized by adding an anaesthetic to the water before the laboratory exercise. Carbon dioxide from a gas-operated wine cork remover is a possible method.

Procedure

1 Students work singly and the samples are divided evenly amongst them.

2 The samples are mixed by gentle stirring to randomize the catch.

3 Organisms are pipetted into developing trays, either along parallel lines drawn on the base of the tray or into the narrow grooves in the base.

4 Whichever technique in 3 is adopted arrange organisms into one or more lines with a paint brush.

5 Organisms are recorded by classifying them into runs of the same species. For example, if a line of organisms of 3 speices, A, B, C were arranged as below,

```
A  B B  A A A  C  B  A A A  C C  B
1   2     3    4  5    6      7  8
```

then there are 14 organisms and 8 runs.

The simplest way to record large data sets is to record 1 for the first organism then 1 for each change that is encountered. For the above example this becomes;

```
A  B B  A A A  C  B  A A A  C C  B = 14
1  1    1      1  1  1      1    1 =  8
```

6 Data are collected until 200+ organisms have been recorded.
7 Calculate the Sequential Comparison Index

$$\text{SCI} = \frac{\text{number of runs}}{\text{number of organisms}}$$

8 Compare results between habitats. Does the total number of species from each habitat, i.e. the species richness, show the same trends as the diversity measure? What are the properties of the habitats with the highest and lowest diversity scores?

Exercise 44
Diversity in woodland trees

Apparatus

100 m tape (or 2 × 50 m); bamboo poles.

Preparation

This exercise requires an accurate identification of all the tree species to be encountered in two adjacent mixed woodlands. To assist the identification, a simple key can be produced based on leaf characters recorded during a preliminary visit. Before the practical, a similar-sized woodland compartment is marked in each of the adjacent woodlands. The size of the compartment depends on the density of trees and number of students. A pair of students can usually cope with a 100 m × 100 m quadrat in each of the two woodland compartments within one afternoon, but this may need adjusting in very dense regenerating areas or sparse open woodland. Allocating a pair of students to each 100 × 100 m area, 18 students can record the total tree inventory of a compartment 300 × 300 m in each of the two woodlands. Both the outline of the large compartment and the 100 × 100 m quadrats should be marked with bamboo canes and numbered prior to the practical to save time. Likewise, the closer the woodlands the better as long as they show some contrast in management or physical environment, e.g. grazed/ungrazed, dry/wet.

Procedure

1 Students work in pairs.
2 Each pair of students is allocated to a quadrat in each of the woodlands.
3 Within the first area students record a list of the species present and the number of each species. Chalk can be used to mark trees to avoid double recording. In addition, general observational notes relating to the vegetation, physical structure of the woodland, leaf litter, soil and light climate are taken.

4 Students move to the second woodland and repeat 3.
5 For each of the two woodlands the class results are pooled.

Woodland compartment A		*Woodland compartment* B	
Species	*Number of individuals*	*Species*	*Number of individuals*
1	N_{A1}	1	N_{B1}
2	N_{A2}	2	N_{B2}
3	N_{A3}	3	N_{B3}
4	N_{A4}	4	N_{B4}
S	N_{AS}	S	N_{BS}
	$\Sigma N_i = N_A$		$\Sigma N_i = N_B$

6 Calculate the diversity for each of the woodlands from

$$H = -\Sigma \frac{N_i}{N} \log \frac{N_i}{N}$$

where N_i is the number of individuals of the ith species and N is the total number of trees.

For woodland compartment A,

$$H_A = -\left\{ \frac{N_{A1}}{N_A} \log \frac{N_{A1}}{N_A} + \frac{N_{A2}}{N_A} \log \frac{N_{A2}}{N_A} + \frac{N_{A3}}{N_A} \log \frac{N_{A3}}{N_A} + \ldots \frac{N_{AS}}{N_A} \log \frac{N_{AS}}{N_A} \right\}$$

7 Compare and contrast the woodland diversity values using the field notes. In many situations the S values, i.e. the numbers of species, are similar but the H values are different. Since in this case the two woodland compartments are fully censused there is no need to calculate standard errors to enable values to be compared. If, on the other hand, we were using small samples to calculate the diversity of the large woodlands as a whole then the process is more complex (see p. 166).

Discussion and conclusions

The Sequential Comparison Index is a useful rapid technique for estimating diversity. It is applicable to many situations where the sample can be randomized (e.g. random distances along a transect through vegetation or across a rocky shore can be used to randomize a sequence of naturally occurring organisms). The diversity values obtained correlate well with those values derived from the Shannon-Wiener Index (Williams and Dussart, 1976). The exercise which examined changes in freshwater invertebrate diversity is a simple laboratory study which enables differences in habitat variables to be related to the diversity of these organisms. In this case the interpretation of the results can only be speculative unless they are combined with a comprehensive water survey or used to make comparisons between flow rate and diversity when the samples are taken from the same river system. When combined with other survey data the Sequential Comparison Index can be used to study quantitative relationships. Plankton can also be studied in a similar way. Freshwater organisms are large and easier to manipulate and have been much used as an index of water purity in pollution studies. Many of these studies (e.g. Williams and Dussart, 1976) use diversity indices of which the Sequential Comparison Index is the most rapid. It is worth noting that the work of Williams and Dussart also includes practical details of the chemical assessment of water purity which would be of use in a more detailed freshwater study.

The use of more complex indices of diversity such as the Shannon-Wiener Index causes a number of problems which relate to the underlying mathematical theory (see Pielou, 1974). In the woodland example diversity is measured by taking a complete census of trees in a compartment and the standard formula for diversity $-\sum N_i/N \log N_i/N$ can be used assuming that all species present are well represented in the sample. Where this is not the case (as with the ground flora) an adaptation known as the Brillouin Index has to be made,

$$H = \frac{1}{N} \log \frac{N!}{N_1!N_2!N_3!\ldots N_S!}$$

If a number of sample quadrats is taken to represent a plant community then the results should be pooled and a Brillouin Index calculated. If the index is calculated for each quadrat separately and the result averaged, a much smaller value than the true value will result since each individual quadrat will be a poor representation of the community (*c.f.* species area curves, p. 6). If the diversity of a number of samples is measured by pooling the contents of successive quadrats, eventually the diversity value (Brillouin Index) becomes constant. Pielou (1974) gives a full account of the problems associated with the calculation of diversity values from samples, and a method for calculating the standard error.

Further investigations

1 It is possible to date hedges by a simple diversity index (Pollard, Hooper and Moore, 1975), where the age of the hedge in years is given by:

$$110 \text{ (no. of woody species in a 30 foot length)} + 30.$$

Here we are using a measure of diversity—in this case the simplest type based on species richness–to measure age, based on the assumption that hedges take a very long time to acquire new woody species. Of course the assumption is not always valid and can only ever offer an approximate measure, but the results of the diversity index technique when compared to those of other techniques of ageing are very encouraging.

Because the method is so simple and rapid it is especially suited to a wide range of field course projects. The age index can be used to compare the flora and fauna of different hedges. The measures of flora and fauna can themselves be associated with different hedges by diversity indices. Similarly, comparisons may be made between hedges of the same age but different constituent tree species.

2 Diversity indices have been much used to study freshwater communities (Wilhm, 1972) and in particular water purity. River surveys using the techniques of Williams and Dussart (1976) are suitable for long term student projects or the combined efforts of many students during intensive field courses.

3 The sub-littoral zone of the seashore is another habitat which can be studied with the help of diversity indices. Exposure to wave action, slope and substrate are all factors which affect the diversity of biota found on any shore. In regions where headlands provide a variety of shores, comparisons of the zonation patterns, the overall diversity and the diversity within each zone are possible without excessive travel. The Field Studies Council (Montford Bridge, Shrewsbury, U.K.) would be pleased to advise on centres suitable for such studies.

4 Plant diversity in grassland can often be directly related to management (Grime, 1974). If a number of grassland types is studied (e.g. parkland, pasture, roadside verges, ornamental lawn) it should be possible to relate management to diversity, although quantifying management is a difficult problem. Grime has provided examples from a wide range of habitats where 'disturbance' and 'stress' factors are examined and compared with the numbers of species able to survive in any particular habitat. The work of Grime is suitable for project work related to both the competitive strategies of individual plants and the classification of vegetation/habitats.

Exercises 45 and 46

χ^2 species interactions

Principles

A useful technique for identifying associations (both positive and negative) between species is to calculate some coefficient based on how often they are found together. The χ^2 statistic is the simplest and most convenient of such measures and allows the calculation of the relationship between many different species occurring together. Since, for large numbers of species, the number of individual calculations involved is very tedious, the method is best used in conjunction with a programmable calculator or computer. The method involves the collection of data in the form of species presence/absence data from a number of sample sites (usually quadrats). A χ^2 calculation is performed between each species and all other species and the extent of their association noted. The χ^2 values for each pair of species are examined for significance by reference to χ^2 tables with one degree of freedom. If the number of times two species occur together is greater than expected by chance they will be positively associated, whereas if they are found together less than expected they are negatively associated. The size of quadrat or other sampling unit is important in this context; see Section 1. Also, the establishment of a significant association between species should not necessarily be interpreted as a causal relationship, only a spatial one.

The next stage in the analysis is to compare all the species' associations at once. This is achieved by two methods; firstly the erection of a two-way table exhibiting all the species associations and secondly by the construction of constellation diagrams, which are a form of reticulate classification. The two-way table is formed by using a half-matrix as in Fig. 45/46.1, which shows the positive associations calculated between species of woodland ground flora in a south Devon wood. Species with no significant associations are ignored. Negative associations may also be entered in the same table. Next, a two-dimensional species constellation is formed by plotting the distances between individuals (as best one can) using the reciprocal of their χ^2 values. In this way species highly positively associated will be positioned close together. It is usual to emphasise relationships by using different colours or boldness of line to denote the level of association, e.g. as in Fig. 45/46.2, where the levels of boldness refer to $p \leqslant 1\%$; $P \leqslant 5\%$. Examples of χ^2 associations

		E.n.	S.d.	O.a.	A.n.	A.f.	H.h.	P.s.	D.d.	U.d.	G.a.	R.f.	A.m.	C.o.	M.p.
E.n.	*Endimion non-scriptus*														
S.d.	*Silene dioica*														
O.a.	*Oxalis acetosella*	++	+												
A.n.	*Anemone nemorosa*	++		++											
A.f.	*Athyrium felix-femina*														
H.h.	*Hedera helix*				++										
P.s.	*Phyllitis scolopendrium*						+								
D.d.	*Dryopteris dilitata*	++													
U.d.	*Urtica dioica*														
G.a.	*Galium aparine*									++					
R.f.	*Ranunculus ficaria*														
A.m.	*Adoxa moschatellina*							+			+	++			
C.o.	*Chrysosplenium oppositifolium*							++		++	++	++			
M.p.	*Mercurialis perennis*									++	++		++	++	
R.b.	*Ranunculus bulbosus*										+			++	+

Fig. 45/46.1 A χ^2 half-matrix showing positive associations between species (+ $= p < 5\%$; ++ $= p < 1\%$).

displayed in this way for large numbers of species can be found in Agnew (1961) and Welch (1960).

Interpretation of constellation diagrams, which can be considered a type of ordination (see Ex. 51), usually relies on the subjective identification of species groupings. However, in most cases the groupings are quite distinct and form realistic plant communities or plant associations, with the advantage that links between groups are displayed and species affinities are readily studied. Some species will only be found in a particular community and possess strong affinities to this, whilst others will be general to two or more associations, and yet others will be found throughout the study area and have no strong association with any other species or groupings. The χ^2 constellation provides evidence of these relationships in a rapid and clear manner and is thus a useful teaching aid in the study of communities.

Exercise 45
χ^2 associations of British birds

Apparatus

BTO Atlas of Breeding Birds of Great Britain and Ireland (Sharrock, 1976); random number tables.

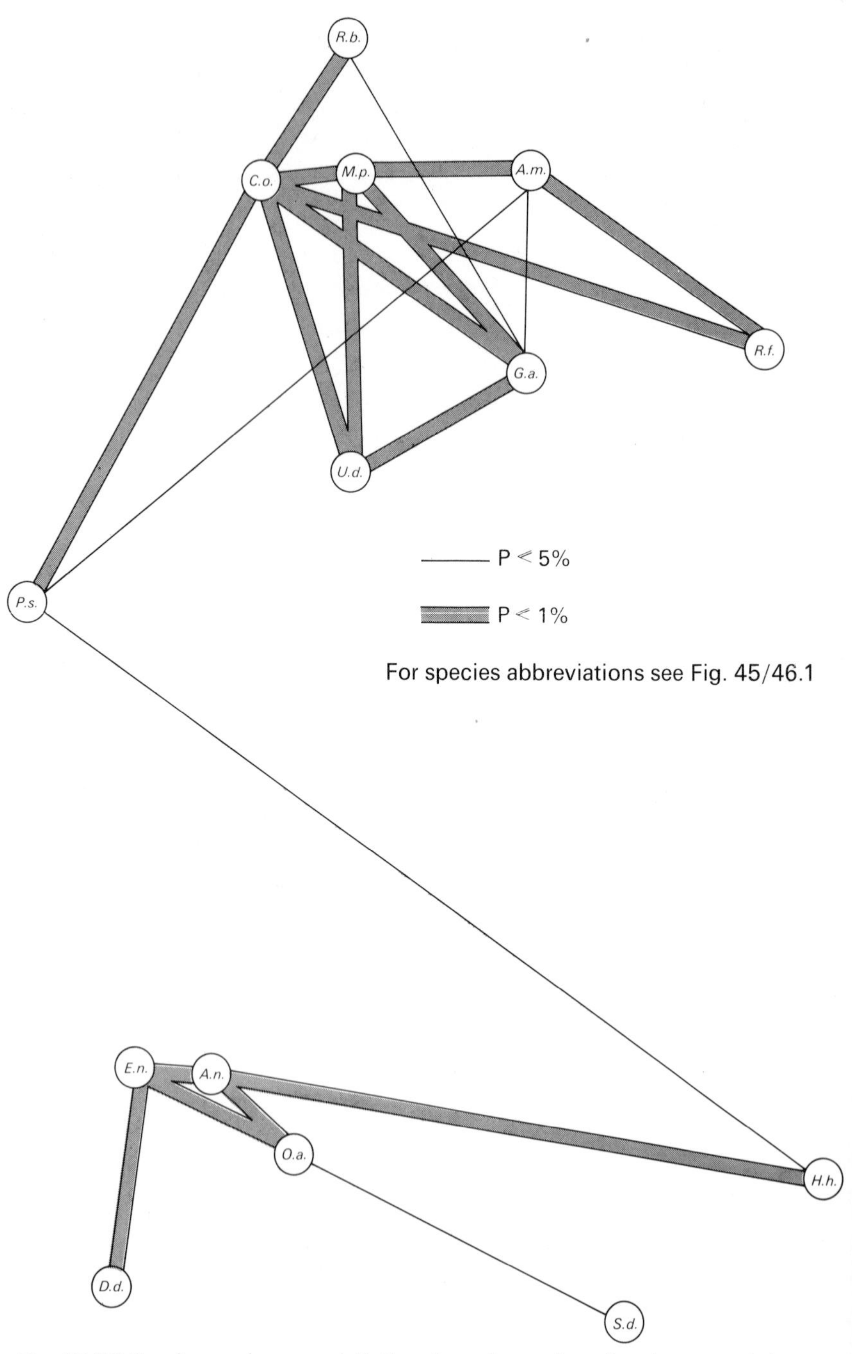

Fig. 45/46.2 A species constellation based on the χ^2 values used in Fig. 45/46.1

Preparation

1 Obtain permission from the publishers to photocopy distribution maps from the BTO Atlas as selected by random number tables; 50 species maps should be sufficient.
2 Random 10 km squares are selected (1 per student) and given an arbitrary number.

Procedure

1 Students work singly, each with a copy of each of the selected maps.
2 Each student is allocated a km square on the maps and works through the distribution maps recording each bird species occurring in the km square.
3 Tabulate class results:

km sq.	*species*					
	1	2	3	4	5	.....
1	√	—	√	√	—	
2	—	—	√	√	—	
3	—	√	—	—	√	
4	√	—	√	√	—	
.						
.						
.						

4 Calculate χ^2 values from a contingency table for each pair of species. For species A and B the calculations are,
a–km squares with both species A and B present
b–km squares with only species B present
c–km squares with only A present
d–km squares containing neither A nor B

$$\chi^2 = \frac{(|ad-bc|-\frac{1}{2}n)^2}{(a+b)(c+d)(a+c)(b+d)}$$ See p. 14

where n = total number of km squares.
5 For the significance level of the results refer to a table of χ^2 with one degree of freedom.
6 Erect a two-way table showing all species associations.
7 Construct a χ^2 species constellation (see p. 168).
8 Identify any species groupings present.
9 Compare the groupings with the distribution of woodland, highland, climatic zones, etc. supplied as overlays with the BTO Atlas.

Exercise 46
χ^2 associations of woodland flora

Apparatus

1 m^2 quadrats; random number tables; compasses; data sheets.

Preparation

1 Select a deciduous woodland offering a number of distinct communities of ground flora. In fact this type of analysis is suitable for the study of many vegetation types. Heathland, lawn, salt marsh, sand dune and bog vegetation have been studied using this technique and the results usefully compared with those from hierarchical classification and ordination techniques.

2 Construct data sheets from a species list for the selected area.

3 To assist interpretation, detailed maps of the study area showing contours of height, pH, soil moisture and light at ground level and other habitat features should be produced. In large groups, it is possible to divide students into two groups, one collecting habitat data and the other collecting vegetation data.

Procedure

1 Students work in pairs.

2 Assuming a large stride is 1 m in length, students use random number tables to provide *X* and *Y* co-ordinates to locate sites throughout the woodland and the compasses to maintain a straight course through the vegetation.

3 At each site students record the vegetation in 1 m^2 quadrats as presence/absence records.

4 Each student pair should aim to complete 5 quadrats.

5 On return to the laboratory (*or* a second afternoon) χ^2 associations must be calculated for every possible permutation of species pairings.

For two species A and B the calculations involved are,
a – quadrats with both species A and B present
b – quadrats with only B present
c – quadrats with only A present
d – quadrats containing neither A nor B

$$\chi^2 = \frac{(|ad-bc|-\frac{1}{2}n)^2}{(a+b)(c+d)(a+c)(b+d)}$$

where n = total number of quadrats.

7 Refer the results to a table of χ^2 with one degree of freedom for significance.

8 Erect a two-way table showing species associations.

9 Construct a χ^2 species constellation.

10 Identify any groupings present and attempt to interpret these groupings with reference to the field situation, or directly with habitat data.

Discussion and conclusions

A χ^2 analysis of associations attempts to identify species associations, not just between single species but in a multivariate way so that clusters (groupings) of associated *species* are identified. This is somewhat different from the results of the classification and ordination methods described in Ex. 52 and Ex. 54 which group *sites* according to their similarity.

For the bird data, groups of birds which are found only in woodland, moorland, mountain or marine habitats tend to form clusters, whilst hedgerow birds and species commensal with man, e.g. the house sparrow, are only loosely linked with all groups. Some groupings will be connected with northern or southern climatic limits of species, and others with no obvious reason, e.g. the crested tit in pine forests of north Scotland. In a similar way local bird count data can be tested for associations of species which are themselves tied to a particular habitat.

Woodland ground flora associations have been used for such comparisons and provided interesting discussion. The highly positive χ^2 associations are usually found between plant species with similar specific narrow habitat requirements, e.g. calcicoles, shade loving plants, etc., and the negative associations between plants with opposing habitat requirements. Most plants however appear, on analysis, to be neutral or weakly positively associated. This phenomenon is related to the size of sample (see Section 1). If too large a quadrat is used all species will be positively associated on analysis, whereas too small a quadrat will produce mostly negative associations. When an appropriate sized quadrat is used (1m × 1m to 4m × 4m is a suitable range of sizes for woodland ground flora) there will be some species which are found in all quadrats, e.g. *Rubus fruticosus* (bramble) and *Hedera helix* (ivy) are often widely distributed in woodlands and form outliers to several species constellations or link two or more of them together.

Further investigations

1 χ^2 associations can be compared with the results of both classification and ordination on data from the same sites. A study of what each method sets out to do and any differences in the results gives a better understanding of vegetation description than any one method on its own.

2 As already mentioned, the associations identified by the χ^2 species constellations do not necessarily suggest causative relationships but may instead relate to an environmental factor controlling the distributions of several species. In some cases this is immediately obvious but in others it

is difficult to establish which of the two factors predominates. Apart from the examples used in Exercises 45 and 46 there are several other ecological situations suitable for analysis; the invertebrate fauna of carrion through time is likely to produce both types of association; also, aquatic macrophytes where depth and water flow rate would be involved and the flora and fauna of rocky shores where exposure is likely to be the dominating distributional factor.

3 Other coefficients such as the variance coefficient (Williams and Varley, 1967) may be used to establish species associations and in a longer term project the effect of the choice of coefficient on the results of the species constellation could be investigated. Such work could be further extended by also using a number of different data sets with different characteristics.

4 Once a positive or negative relationship is suspected from a species constellation then this may be investigated by correlation analysis. Both the relationship between species and between species and habitat factors can thus be studied. Usually the process involves the collection of new data, this time in a quantitative form, from random or systematic samples. The values for the various species or habitat variables are then correlated (see Parker, 1979) and the significance tested.

5 Species interactions can be analysed using other floral or faunal atlases, which exist, for birds, for an increasing number of European countries. In Britain, the floral atlas of the Botanical Society of the British Isles would also be appropriate.

Exercises 47 and 48

Distribution and abundance of organisms in time and space

Principles

If a community of animals is sampled regularly over a period of time, some species may occur in most of the samples while others may occur on a few occasions only. The total number of animals of a particular species will also vary, ranging from very common species to very rare ones. Using their occurrence in time we can classify the species in a community sampled in this way as regular, not regular, irregular or seldom according to the frequency of their appearance. With reference to the total number within each species, we can use a classification such as rare, not common, common or abundant. The two classifications together can therefore locate each species on a graph where abundance is the ordinate and frequency the abscissa.

Instead of classifying the distribution of animals in time, we can consider their distribution (or that of plants) in space. We would then group the records on the graph in categories such as widespread, i.e. occurring in most of the sample units (quadrats, lakes, trees etc.), not widespread, local or very local, retaining the abundant/rare classification for describing the total numbers of organisms found. A species which falls in the category 'abundant but local', for instance, would immediately draw the attention of a conservationist; because this classification takes account of total numbers *and* spatial distribution, it shows that the above species (and perhaps its habitat) may require active management to ensure its conservation. The species is local even though it is abundant where it occurs. Williams (1964) has classified the spatial characteristics of water beetles in 13 Canadian lakes in this way, and used temporal data from bird counts made during a series of visits to one area to classify the birds using the abundance/regularity method. In both methods (i.e. space and time) a line depicting a random distribution can be calculated. For any number of individuals this is the arrangement of them among the sample units (time or space) which would be expected by chance causes (Fig. 47/48.1). An animal of a particular abundance category which falls to the left of this line is therefore aggregated among the sample units, while one to the right of the line tends towards maximum dispersion (to the limit of one individual per sample unit). To plot the line of random distribution, the expected number of units *unoccupied* is calculated from the formula for the Poisson distribution for

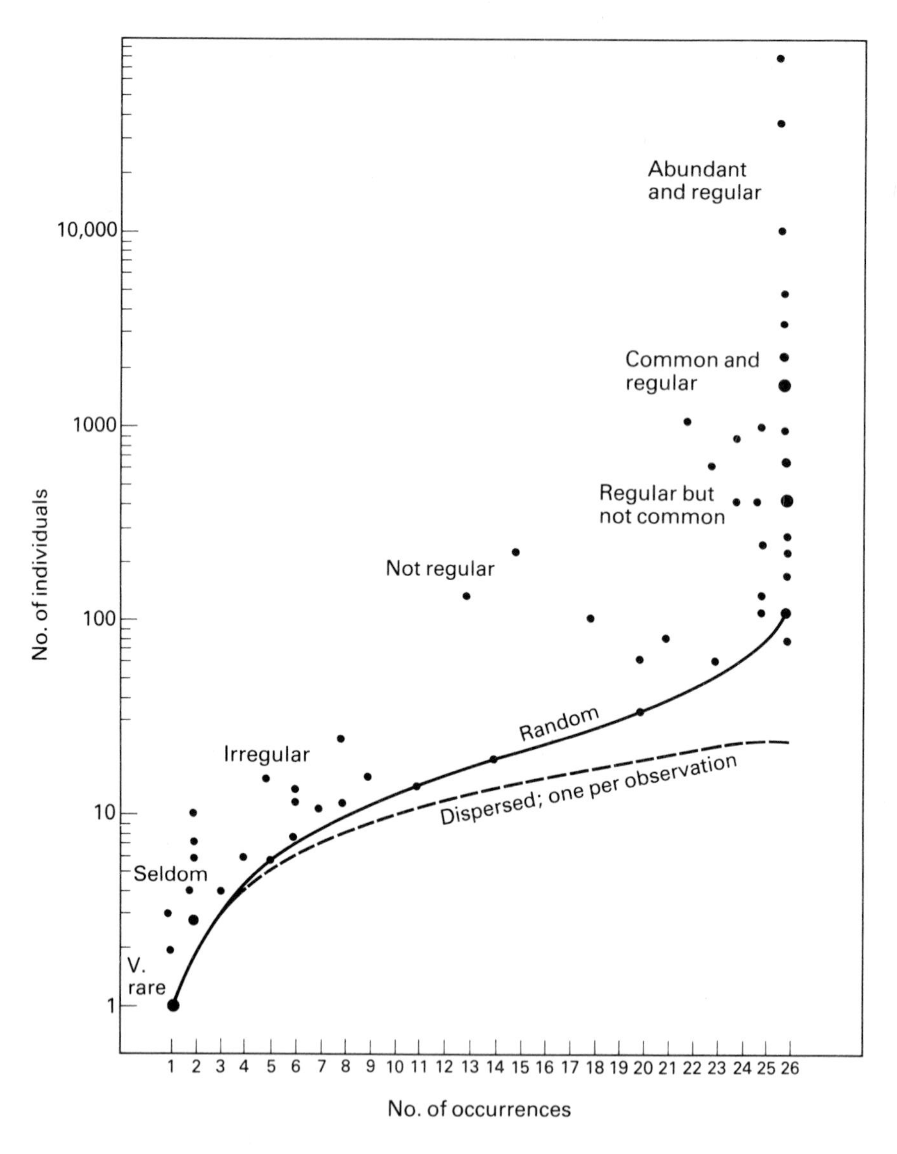

Fig. 47/48.1 The relationship between the number of individuals recorded and the number of sampling occasions on which each species was recorded (after Williams, 1964).

a range of numbers of individuals up to the point where the line becomes vertical. The number of occupied units is therefore calculated by subtraction (see p. 54 for the formula for the first (zero) term of the Poisson series). For example, with 100 sites available for occupation and

20 individuals to occupy them, according to the Poisson series the mean number/site is 0.2. This figure is used in the formula

$$n.e^{-m} \text{ where in this case } n = 100 \text{ and } m = 0.2.$$

So in this case, the number of sites out of 100 unoccupied is $100 \times 0.82 = 82$ sites. Therefore, the first point we can plot on the graph is for 20 individuals occupying 18 sites.

In the laboratory exercise which follows, the species in a time-series of light-trap collections of moths are classified according to their abundance and temporal distribution, while the field exercise considers the abundance and spatial distribution of insects captured in different parts of a habitat.

Exercise 47
Abundance and regularity of appearance of species in a moth community

Apparatus

A light-trap (mercury vapour or tungsten lamp—see Southwood, 1971); reference collections of moths (see below); naphthalene balls; killing jar with tissue paper and ethyl acetate. Polystyrene cups with lids—three times as many as there are students. A copy of South (1961).

Preparation

1 Operate the trap in a rural or semi-rural area for a number of consecutive nights during the spring, summer or autumn. Examine the catch each morning and either:

(*a*) kill all the moths and take them back to the laboratory/classroom;

or (*b*) identify the moths without killing them and record the number of each species. Make a reference collection of one example of every species captured during the catching period and make a note of the night's catch from which they were removed.

2 If (*a*) above is followed, the number of nights for which the trap should operate should be determined by the number of students taking part and the number of insects each student can deal with in the time available. Using the reference collection, a student can identify *c.* 100 moths in an hour. If some nights' catches are lower than the required minimum, more than one night's moths (in separate containers) should be allocated to a student; if more than the required number occur on one night split the catch into separate containers, give these to different students, but pool the data for that night. Store the moths (after oven-drying at *c.* 40°C) in the dated, polystyrene cups, each with a naphthalene ball and a lid. Alternatively, pin each night's catch on a polystyrene tile.

3 For a small class, especially one where this exercise is combined with another, related one, the number of moths required is not excessive and should not significantly reduce a local population. If desired, the rarer species only can be released, or, for the more abundant species, a known proportion of the catch each night can be retained and a record kept of the number released. Whatever method of collecting/recording is adopted, have available on the day of the exercise reference collections of the moths captured (for instance, a box for rarities and another for commoner species) and provide each student with the data or the actual catch for one or more nights.

4 Prepare a blackboard tabulation in alphabetical order of all the species known to have been captured. Leave space opposite each name so that each student can enter the number of that species captured on their allocated date. The tabulation does not have to include the actual dates; it need only show how many moths of each species were caught and on how many different dates. If the latter are included, however, much more information is available on flight periods etc.

5 Knowing the total number caught for the most abundant species over the catching period, calculate the random distribution curve from the Poisson distribution for a range of numbers from this maximum down to one individual. Before the practical class a graph should be prepared with this curve and that for maximum dispersion already drawn (i.e. 10 animals occurring on 10 dates, 20 on 20 and so on). During the exercise, students can be allocated moth species to plot on the class graph. The graph will later receive a pinned specimen of every moth species caught so its size will be determined by the number of species.

Procedure

1 Each student should arrange their allocated moths in piles of similar species and identify each species from the reference collection; if the moths are pinned, they should be rearranged in groups of species.

2 Enter on the blackboard for each species of moth the number of individuals captured on each date.

3 When all the data have been entered, nominated students should extract the totals for one or more species and plot them (no. of individuals and no. of nights captured) on the class graph.

4 Pin an example of each species at the appropriate position on the graph, each with its name on a label on the pin or written on the graph.

5 Draw lines round groups of species of similar abundance/regularity classifications, paying attention especially to those which deviate from the random line. Discuss the implications concerning flight periods etc. of those species which lie well above the line or between it and the regular dispersion line.

Exercise 48
Abundance and distribution of insects in a field community

Apparatus

Per pair of students: ten pitfall traps (see Exercise 14); ten canes; at least 16 specimen tubes (2.5 cm × 5 cm) with lids; one pooter and one sweep net (see Southwood, 1971) (a metre-square quadrat may be substituted for the sweep net); a pair of fine forceps. At least 20 white polystyrene ceiling tiles *c.* 30 cm × 30 cm labelled individually with major invertebrate taxa expected to be captured (e.g. a tile for Carabidae, a tile for Heteroptera etc. as appropriate). Prepare two series of tiles, one labelled 'pitfalls' and the other 'sweeps'.

Preparation

1 If the exercise is to be carried out as part of a residential field course, during which students can set up their own pitfall traps on the day before the exercise (see 2 below), the only preparation is the selection of a field with a wide range of plant species.

2 If it is not possible for the participants to set their own traps, set them out 24 hours before the exercise in a grid, the dimensions of which should give as great a coverage of the chosen area as possible, including sites near its boundaries. Mark each trap with a numbered cane.

Procedure

1 Allocate each student a number.

2 Using the pooter and forceps, remove the invertebrate catch from each pitfall trap and put each trap's contents in a separate specimen tube together with the trap's number in pencil on a piece of paper.

3 One member of a pair should sweep the vegetation a pre-arranged number of times (determined by trial sweeps before the exercise, and the time available for identification after the field collection). The catch should be removed with a pooter and placed in one or more specimen tubes (labelled with: sweep 1, sweep 2 etc.). Make sure that not more than one student sweeps a particular area.

4 Repeat 2 with the second member of the pair.

5 As an alternative to 3, each student can search a square metre of vegetation for 10 minutes, removing invertebrates with a pooter or forceps as they are found.

6 On the bench top next to each 'pitfall' tile place a piece of paper with the same number of columns as there are pitfalls. Pin one example of each 'species' or organism captured in the pitfalls on the tile allocated to its taxonomic group. Repeat this for all the tiles (i.e. all the taxonomic groups).

7 Each student pair should sort their pitfall catches and enter in the appropriate column of the class data sheet the number of each type of organism captured (e.g. Carabidae, species A:17; Carabidae, species B:6 etc.), opposite the appropriate specimen on the tile.

8 Repeat 7, using different tiles, for the sweep net catches or surface searching data.

9 Record, for the pitfall and sweep net data separately, the number of individuals captured of each invertebrate 'species' pinned to the tiles and the number of sample units in which they occurred (i.e. the number of pitfall traps or the number of student sweep series; the maximum value for the latter equals the number of students).

10 Allocate each species a number and construct two class graphs of abundance/distribution; one for the sweep catches/searching and the other for the pitfalls. Calculate the line for random distribution and plot this and that for extreme dispersal on the class graphs.

11 Allocate the 'species' to the students who should plot each species on the class graphs. Remove the pinned invertebrates from the tiles and pin them in the appropriate position on the graph. Draw lines round groups of similarly-classified animals.

Discussion and conclusions

Both the field and laboratory exercises should show how Williams' method for classifying the members of a community is a useful way of arranging large amounts of ecological data. It could be used in many situations where long-term or wide-ranging surveys of plants or animals have already been made but have not been analysed. The moth exercise should show how the catch is dominated by one or two species which fly for most of the period. The longer the catching period, however, the less likely is it that one or more species will occur in the regular category. If a whole year's data were treated in this way it is doubtful if any single species of moth would be caught on most days, so the plotted data would be clustered near the y axis. In Britain, 40–60 consecutive nights' captures have been found to give a good spread. Reference to details of the moths' life history and food plants in South (1961) should provide clues to the reasons for individual species occurring where they do on the graph. A common species with a wide food-plant range might be spread over a long catching period while a rare migrant may occur in the trapping area once only; the latter distribution raises problems relating to the definition of the community. Consideration of the life cycle and feeding habits of the invertebrates captured in the field exercise may also partly explain their classification; local and rare species probably have specialized feeding habits or, like the migrant moths, may not even be part of the local fauna.

Further investigations

1 The nightly moth catches in Exercise 47 can easily be analysed using the diversity index of Cairns *et al.* described in Exercise 43. If all the moths were not killed, students can be given as many squares of paper as there were insects in that night's catch. The pieces of paper should be given numbers representing the species and they can be randomly arranged on the bench top to provide the index. Changes in the diversity index through the catching period could be related to weather or the presence or absence of abundant species.

2 If meteorological records are taken each night of the moth-trapping period, analyses (such as multiple regression) can be attempted relating numbers captured to temperature etc. Exercise 28 could be modified to determine temperature thresholds for flight in selected species.

3 To demonstrate the abundance/rarity method in the field exercise it was not necessary to identify the captures to species level. The quadrat method could easily be used for plants, in which case species identification should be easier but separation of individuals would be more difficult.

4 Instead of using a time-series of moth captures in the laboratory exercise, animal distribution maps could be used to construct an abundance/rarity classification of the chosen animal group. The Atlas of Breeding Birds in Britain and Ireland (Sharrock, 1976) contains maps of all breeding species plotted on a presence or absence basis in 10-kilometre squares. The accompanying text gives fairly precise information on the estimated *number* of breeding pairs of each species in Britain. All or a selection of the maps could be photocopied (with the publishers' permission) and allocated to students. On the back of each photocopy write the estimated population. The students could count the dots (all size classes together—short-cuts can be devised to make the counting easier) and plot their species on the class graph. Alternatively, the number of occupied squares can be written on the top of the map—it is given in the species accounts in the atlas. This exercise has the advantage that it confirms the classification of some species which would be expected from general knowledge of the group, e.g. gannet (*Sula bassana*)—locally abundant, because it nests in very large numbers at a few coastal sites.

5 School, college or county natural history societies often have an accumulation of data on the abundance, distribution and regularity of appearance of plants or animals, and the above methods could be used in some of these cases.

Exercises 49 and 50

Detection of mortality in populations of a leaf-mining moth and a snail

Principles

We rarely see direct evidence of organisms dying in the field and usually have to analyse indirect evidence. A common method is to collect this indirect evidence in the form of a time-specific (Exercise 19) or age-specific life table. The latter is essentially a series of counts of the numbers of organisms through a generation. In a moth, for example, the fate of a cohort of eggs, laid more or less at the same time, would be followed through to the emergence of adult moths, the number and cause of deaths being recorded at stages throughout the insects' development. The construction of life tables of this sort usually assumes that different mortalities are acting successively. The 'strength' of each can then be determined by subtracting the number of organisms remaining after the mortality factor has acted from the number present before it acted. However, the assumption of successively acting mortalities is not always valid. For instance, if parasites killed ten insect eggs in a batch of 200 and, *during the same period*, 15 eggs were eaten by predators, the subtraction process described above would be difficult because we would not know from which egg total to subtract. The life table in Exercise 49 'overcomes' these problems in the usual way (i.e. assumes mortalities act successively). Like most other life tables, it also suffers from the disadvantage that, for reasons of sampling or availability of the organism, we do not have mortality data on all stages of development. Any fundamental information we obtain concerning the main mortality factors acting on the organism obviously relates only to those developmental stages or mortality factors we have investigated.

The laboratory exercise concerns a moth, the larvae of which mine the leaves of holm oak (*Quercus ilex*). Leaf mines are useful for exercises on life table construction because much evidence of mortalities (parasite pupal cases, bird damage etc.) remains in or on the mine at the end of the generation. The usual gaps in the data occur, however, and the main omission in this life table is that we have no estimate of the number of eggs laid; our starting density is given by the number of mines/thousand leaves, i.e. the number of successful penetrations by young larvae.

For each mined leaf in turn, we use a key to determine whether the moth reached the adult stage or whether it died as a young or old larva. This is facilitated by the fact that the fourth instar larva spins silk in the mine which, on drying, contracts and pulls the mine surface (abaxial leaf surface) into a longitudinal ridge. This gives the larva more room to feed and grow and tells us that at least the fourth instar was reached. Larvae are commonly predated by birds (mainly tits, *Parus* spp.) in Britain, and a jagged tear in the mine surface signifies this. Among other causes of mortality are various parasitic wasps; the key includes only those which are most common in Britain. A preliminary examination of a collection of mines would be useful; any extra parasites or other mortality factors could be identified and need only be designated A, B, C etc. and the class told the criteria used to separate them. Because of the obvious limitations, the construction of the leaf miner life table is very much an exercise in life table construction rather than in the accurate quantification of *Phyllonorycter* mortalities. Clear density relationships often result, however, and because the method uses data collected by the class the exercise probably teaches more about life table construction than does the use of data borrowed from the literature.

The conversion of the data to 'killing powers' (k-values) is also rare in class exercises and has been found to be a useful way to teach the mechanics of k-factor analysis. Details of its principles may be found in Southwood (1971), Varley, Gradwell and Hassell (1973), and Dempster (1975).

Detection of mortality in a free-living field population is attempted in Exercise 50. This exercise again uses indirect evidence of mortality and compares the frequency of genetically-determined colour morphs in juveniles and adults of a common snail from different areas. We look for evidence that snails of a particular pattern suffer greater mortality than other groups, i.e. that selection is occurring.

Exercise 49
k-factor analysis of mortality in the holm oak leaf miner *Phyllonorycter messaniella*

Apparatus

Per pair of students: up to 200 mined leaves (see below); one pair of fine forceps; prepared key to causes of mortality (see below); a binocular microscope.

Preparation

1 At the end of April/early May (in Britain) collect mined leaves from at least 6 holm oak trees (*Quercus ilex*) to provide the required number for the practical. This is at the end of the miner's life cycle. For each tree in turn, count and record the *total* number of leaves on a small branch, then remove the leaves with mines and count the number of mines; there are sometimes several on one leaf. Do this for a series of small branches

to provide an average figure for the number of mines per thousand leaves for each tree. Obtain the widest possible density range between the trees.

2 Unless the practical is due in a few days, carefully put the leaves in polythene bags labelled with the tree from which they came and put them in a deep freeze until needed.

3 Prepare one copy/student of the following simplified key to causes of mortality:

Key to determine the causes of mortality in the holm oak leaf miner, Phyllonorycter messaniella.

---

1. Mine with ridge: 2
- Mine without ridge: 7

2. Mine surface intact apart from pupal case of moth projecting from emergence hole. *Adult stage of moth reached*
- Mine surface intact apart from pupal case projecting and dead moth stuck in case. *Death of adult—accident on emergence*
- Mine surface intact, no holes, no pupal case projecting: 3
- Mine surface *not* intact; a jagged tear *or* a *small, neat* round hole: 6

3. Inside mine (remove base carefully with forceps): dead larva, no evidence of parasites (i.e. no pupal cases etc.). *Death of larva (IV+) from unknown causes*
- Inside mine: little evidence of larva or parasites; frass (faeces) and a few moulted head-capsules only *Death of larva (IV+) from unknown causes*
- Inside mine: larval remains present or absent, but in either case a parasitic wasp's pupa (or empty pupal case) present: 4

4. Pupa or pupal case in a whitish silk coccoon looking like:  either intact or with a neat round hole in coccoon wall. *Death (IV+) due to the parasite* Apanteles circumscriptus
- Pupal case in coccoon looking like: but *ragged* hole in side. *Death (IV+) due to* A. circumscriptus *but* Apanteles *then hyperparasitized by* Sympiesis
- Pupa or pupal case black or brown, free in mine (*not* in coccoon): 5

5. Pupa or pupal case jet black. *Death (IV+) due to the wasp* Chrysocharis gemma
- Pupa or pupal case brown. *Death (IV+) due to the wasp* Sympiesis sericeicornis

6. Neat, round hole in mine surface: 4
- Large, jagged hole in surface; no larvae inside; frass and head-capsules only. *Death (IV+) due to bird predation (probably by tits—*Parus *spp.)*

(continued)

7. Mine surface intact: 8
– Mine surface with large, jagged tear; no larva inside.
Death (I–III) due to bird predation (probably by tits—Parus *spp.)*
8. Inside mine: larva dead or no trace of larva other than frass and moulted head-capsules; no evidence of parasite pupae.
Death (I–III) from unknown causes
– Not like any of the above categories. *Fate unknown*

4 Prepare a blackboard tabulation for the class life table for each tree as follows

Mortality category	*No. mines in category–individual students' data*	*Class totals*	*Corrected* class totals*	*Nos. surviving action of mortality factor*	*log nos. surviving (2 decimal places)*	k
No. mines examined	$a_1\ a_2 \ldots a_i$	A	α	α	$\log \alpha$	$\log \alpha - \log(\alpha-\beta) = k_1$
Death of larvae I–III, unknown causes	$b_1\ b_2 \ldots b_i$	B	β	$\alpha-\beta$	$\log(\alpha-\beta)$	$\log(\alpha-\beta) - \log(\alpha-\beta-\gamma) = k_2$
Death of I–III, birds	$c_1\ c_2 \ldots c_i$	C	γ	$\alpha-\beta-\gamma$	$\log(\alpha-\beta-\gamma)$	etc.
Death of IV+, birds	etc.	etc.	etc.	etc.	etc.	
Death of IV+, *Sympiesis*						
Death of IV+, *Apanteles*, including hyperparasitism category						
Death of IV+, *Chrysocharis*						
Death of adult on emergence						
Adults emerged						

Overall generation mortality, $K = k_1 + k_2$ etc.

Procedure

1 For each mine in turn, use the key to decide at which stage mortality occurred or whether the moth reached the adult stage. When a student has examined all the allocated mines, the total number in each mortality category (a_i, b_i, c_i etc.) should be entered in column 2 on the blackboard.

*2 Enter for 'no. of mines examined' in the corrected class totals column the recorded field density of mines/1000 leaves for that tree (α). Then correct the class totals for each mortality (B, C, etc.) by this ratio, i.e.

$$\beta = \frac{\alpha}{A} \times B$$

$$\gamma = \frac{\alpha}{A} \times C \quad \text{etc.}$$

3 When all the data have been collated and the logs calculated, subtract successive logs to obtain the k-value for each mortality on each tree.

4 Add up the k-values to obtain total generation mortality (K). For each tree in turn, draw a circle in brightly-coloured chalk around the largest k-value, i.e. the mortality contributing most to total generation mortality.

5 Compare the blocks of class data to see if the circled k-value falls in the same mortality category on most of the trees. If it does, and if we assume that the pattern between trees within one year is likely to be consistent between years we can designate the ringed k-value 'the key factor'.

6 On a different blackboard or on graph paper, plot, for each mortality category, the k-values for all trees against the log of the density on which it acted; the k_1-values, for example, should be plotted against log α and the k_2-values should be plotted against log $(\alpha - \beta)$.

7 Look for evidence that one or more mortality factors might be acting in a density dependent, density independent or inverse density dependent way, i.e. whether the line through the k-values has a positive, zero or negative slope. Rigorous tests for density dependence in data of this type are given in Dempster (1975) and Varley, Gradwell and Hassell (1973).

Exercise 50
Does mortality rate in snail differ with shell colour?

Apparatus

Per pair of students: 20–30 polythene bags (*c.* 30 × 15 cm).

Preparation

1 Find an area where one or more snail species are common. Ideal species would be *Cepaea nemoralis* or *C. hortensis* which may be found in large numbers at the top of sandy beaches. These species also occur in woodland, where a third species, *Hygromia striolata*, also occurs. All these species exhibit colour polymorphisms and can be identified using Kerney and Cameron (1979). Locate areas where the snails are living on different plant species or different background colours.

2 Carry out a pilot survey of the snails and decide on a more or less arbitrary size division between 'juveniles' and 'adults'. A width of *c.* 8 mm may be appropriate. Decide on a colour classification to be used during the analysis, e.g. banded/unbanded or dark bands/light bands etc.

Procedure

1 Divide the task of collecting the snails between all the student pairs so that from each habitat type (plant species 1, plant species 2, etc.) at least 200 snails are collected. Students should collect systematically from each habitat, i.e. remove all the snails from a small area, to avoid a non-random selection of sizes. Put the snails in bags with pencil-written labels stating the habitat type from which they were collected.

2 Return to the laboratory. Each pair of students should be allocated snails to analyse so that the total number collected is distributed more or less evenly among the class.

3 Each pair should divide the snails into age groups (juvenile/adult) and each age group should be divided according to the light/dark classification already agreed.

4 Pool the class data for each habitat. χ^2 analyses should be carried out for each habitat's data by a group of students as follows:

Habitat 1

	No. banded	*No. unbanded*	*Total*
Adult	a	b	$a+b$
Juvenile	c	d	$c+d$
Total	$a+c$	$b+d$	n

Use the test for a 2×2 contingency table, incorporating Yates' correction to allow for low expected numbers in a cell of the table; the vertical lines mean the absolute, that is, positive, value of $ad-bc$ should be taken.

$$\chi^2 = \frac{n\{|ad-bc|-\frac{1}{2}n\}^2}{(a+b)(c+d)(a+c)(b+d)}$$

If the value for χ^2 is greater than that indicated in the table in Parker (1979) at the 5 per cent level with one degree of freedom, then we can say that the proportion of banded/unbanded snails differs between juveniles and adults in the habitat investigated.

5 After all the data have been collected the snails should be returned to their habitat unharmed.

Discussion and conclusions

Although most life tables do not measure every mortality, as in Exercise 49, *k*-factor analysis is useful in that it shows that often *one* major mortality is the main cause of population change. This mortality is called the 'key factor'. Ideally, life tables are constructed for a large number of successive generations and each *k* is plotted chronologically together with *K*, overall generation mortality. The graphs are then compared visually and the key factor identified as the one most clearly correlated with the fluctuations in *K*. Exercise 49 studies one generation only, so we look for circumstantial evidence for the key factor by comparing trees, not successive generations. When this exercise has been carried out in Britain a clear key factor has emerged, that of predation by birds on large larvae (instar IV onwards). When this *k*-value was plotted against the log of the density, the relationship was inversely density dependent, which is one of the ways one would expect a key factor to operate, i.e. killing a greater proportion at low prey/host densities than at high ones. Alternatively, we would expect the mortality exerted by a key factor to be unrelated to density. We would not usually expect it to act in a density-dependent way because a key factor is responsible for population fluctuations, not potential regulation (see Dempster, 1975).

Other *k*-values may reveal some density-dependence but to be able to decide which factor, if any, is regulating the moth population would require a series of much more complete life tables. See Varley, Gradwell and Hassell (1973) for the next stage of the analysis. Consider also whether the order of mortalities is realistic; the *k*-values could be re-calculated with the egg-laying parasites responding to the density of young moth larvae, for instance.

In the field exercise, circumstantial evidence should be obtained indicating that in some habitats snails of a particular shell colour are more likely to die than animals of another colour. It is difficult to identify the mortality factor however. There are at least four possible explanations of such a result. Young snails destined to become dark banded as adults could develop such colouration late in life, so that classification of juveniles may have been inaccurate. Secondly, selection against one colour morph could be taking place by predators taking more individuals than would be expected by chance. Comparison of shell colour with background should give a clue to this. Thirdly, secondary effects related to shell colour could be occurring. For instance, different coloured snails absorb incident radiation in different amounts and this could affect survival at high or low temperatures. Fourthly, there may be physiological attributes of the snails which are inherited with shell colour and which affect survival in their own right. In the last three cases, we assume that the reason for the observed differences is related to the different action of a mortality factor on different populations, in all of which shell colour is determined by the same genetic mechanism.

Whichever of the above explanations is correct, the exercise shows that the genetical constitution of a population can affect the action of mortality factors so that the 'average' picture given by a life table often obscures more detailed relationships.

Further investigations

1 Exercise 49 could be improved if an estimate of the miners' egg density on each tree could be made available to the class. In Britain, eggs are laid on holm oak from October to December.

2 If the leaf miner data were retained, and the practical repeated in several successive years using the same trees, a plot of each *k*-value and of overall generation mortality against time becomes feasible.

3 Other leaf miners could provide similar life table data. For instance, Lewis and Taylor (1967) give details of the life history of a fly leaf miner on holly (*Ilex aquifolium*).

4 The snail exercise (No. 50) could easily be extended to answer some of the points in the discussion. If adult banded snails are examined carefully, it should be possible to detect the banding in the 'juvenile' whorls, indicating that the first explanation in the discussion is not valid.

5 Longer-term studies could investigate the responses of predators (e.g. birds) to populations of snails whose colour proportions have been determined experimentally, perhaps using the techniques in Exercise 42.

6 Survival of dark and light individuals could be compared by confining a mixed population to an area where temperature extremes are likely to occur but to which predators are denied access.

Exercises 51 and 52

Ordination

Principles

Ordination of data is a mathematical treatment designed to produce an arrangement of sample units (usually quadrats) which reflects their similarity by using a spatial display of their relationships, such that the most similar are closest together in the display. This technique is different from both classification, which conceptually accepts that communities consist of discrete entities (which is often more convenient) and the χ^2 species associations, which examine the species as opposed to the sites. Ordination techniques analyse the similarity between stands and make no assumptions about the integration of vegetation into discrete groups since the arrangement of the data in space places each stand on a continuum. The space is then examined to establish the existence of any discrete groupings which may reflect habitat preferences. However, classification has the advantage of producing results directly amenable to vegetation mapping.

A simple method of ordination based on that devised by Bray and Curtis (1957) is described in these exercises and other methods can be found in Kershaw (1973), Shimwell (1971) and Chapman (1976). In this version of ordination a similarity coefficient is calculated between each stand (sample) and every other stand:

$C = 2W/(A+B)$

where $A =$ the sum of quantitative values for all species in stand A,

$B =$ the sum of quantitative values for all species in stand B,

$W =$ the sum of lesser values for species common to the two stands.

The coefficient is then converted to a dissimilarity index: $D = (1-C)$. These values are then entered into a table comparing each stand with all other stands. The two most dissimilar stands set the end points of the x axis.

If we term the two stands (quadrats) with the highest dissimilarity index between them Q_1 and Q_2, then the position of all other stands can be plotted by referring to the distance of each stand from Q_1 and Q_2, found from the similarity table (see Fig. 51/52.1). The resulting graph is a crude ordination of the species data and the analysis may be terminated at this point or a second axis may be constructed. The second axis is drawn at right angles to the first by selecting two stands which are close

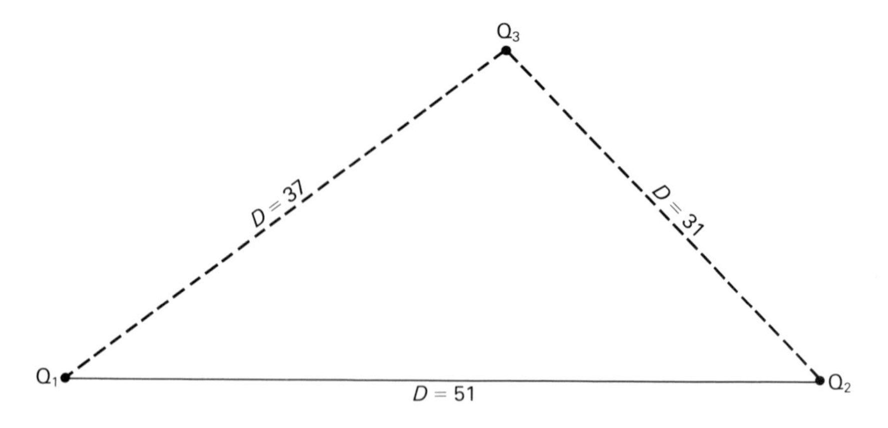

Fig. 51/52.1 Construction of the *x* axis of an ordination showing how the position of a stand Q_3 is set by the distances $Q_1 - Q_3$ and $Q_2 - Q_3$.

to the centre of the first axis but have the largest inter-stand distance (D) in the table, i.e. their D values are far apart even though they are more or less equidistant on the first axis from Q. These two stands Q_3 and Q_4 are the reference points for the second axis (see Fig. 51/52.2).

Once the second axis has been erected (y axis), all other stands can be located with reference to their D values with respect to Q_3 and Q_4. It will be obvious that for each stand there will be two geometrically possible locations, one to the left of $Q_3 - Q_4$ and one to the right. Reference to the distances of each stand from Q_1 and Q_2 will determine the correct one of the two possible locations.

Both calculation of the dissimilarity measure and the erection of the ordination can be programmed for computer analysis.

NOTE: As each site is located on an ordination it must be labelled to allow subsequent interpretation.

Exercise 51
Ordination of freshwater invertebrates

Apparatus

Teat pipettes; white photographic developing trays; sweep nets or freshwater collecting nets; 500 ml plastic beakers with lids; oxygen meter or apparatus for the Winkler oxygen determination method (see Dowdeswell, 1959).

Preparation

1 Before the practical a freshwater habitat, preferably a 'pure' river with fast rapids and slow glides, is selected, thoroughly sampled and the

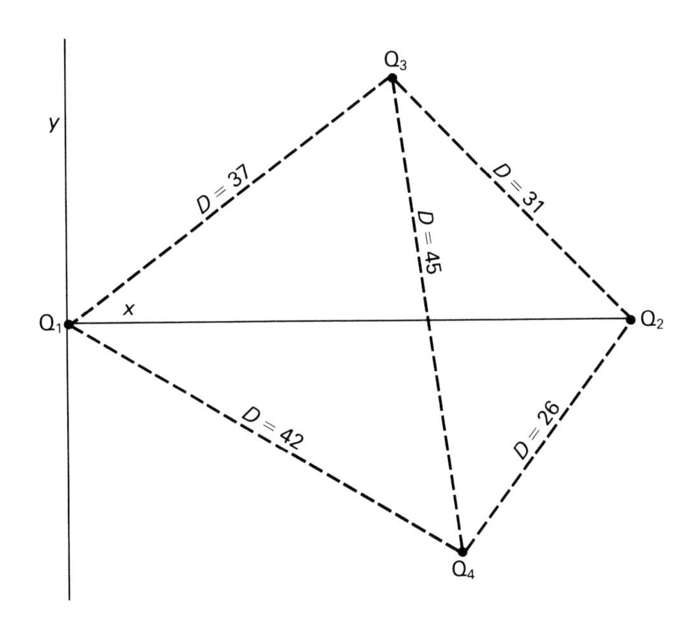

Fig. 51/52.2 Construction of the y axis of an ordination: a stand Q_4 has a position close to Q_3 on the x axis but differs from it by 45 units in the matrix of D values, and therefore should be on the opposite side of the x axis. The selection of which point is represented above and which below the x axis is arbitrary.

organisms identified. From the species list a simple key is produced (visual matching with particular anatomical features emphasized usually suffices). This is to aid identification in the laboratory practical where students will not necessarily need to know the correct species name of each organism (as long as all the students give the same organism the same name or code letter).

2 Duplicate data sheets.

3 Collect enough samples containing freshwater organisms to provide one per student pair. These should be collected as a fixed number of standardized sweeps in various parts of the river.

4 As each sample is collected it is placed in a plastic beaker and labelled.

5 At each station record (*a*) flow rate
(*b*) dissolved oxygen
(*c*) weed cover
(*d*) depth.

6 Ideally the samples should be collected immediately prior to the practical; if this is not possible they must be kept cool (10°C) and oxygenated.

Procedure

1 Students work in pairs.
2 Each sample is allocated to a pair of students who empty it into a plastic tray and return each organism to the beaker having identified it.
3 The data are recorded as the number of each organism in each sample (site) as below.

Site	*Species*				
	1	*2*	*3*	*4*	*5*
1	6	0	3	2	0
2	4	1	0	1	6
3	0	3	10	1	12
4	8	1	6	4	0
5	9	3	8	3	0
6	2	0	0	0	8
7	etc.				
8					

4 Similarity indices are calculated between all pairs of sites from the formula $C = 2W/(A+B)$
5 Dissimilarity indices are calculated from $D = (1-C)$.
For the above example the calculations for the sample sites 1 and 2 would be:

Sample	*Species*				
	1	*2*	*3*	*4*	*5*
1	6	0	3	2	0
2	4	1	0	1	6

$A = 11$
$B = 12$
$W = 4+1$, i.e. the sum of the lesser values of common species.
$C = 2(4+1)/(11+12) = 10/23 = 0.4\,3$
$D = 1-0.43 = 0.57$
$100D = 57$

6 A half matrix of D values is constructed exhibiting all possible sample pairings, i.e. 1 with 2, 1 with 3, 1 with 4, etc. The D values are

often made into whole numbers, e.g. the half matrix of D values can be constructed by using the original D values multiplied by 100.

7 The site pairing with the largest D value, i.e. the most dissimilar, is used to construct the x axis. For a very simple ordination, the analysis can be terminated at this point.

8 A y axis is erected as described on p. 192 and all the remaining stands are plotted and labelled with reference to the two axes.

9 From the completed ordination, identify obvious spatial groupings of D values in the two dimensions of the plot.

10 Separate overlays, e.g. tracing paper or overhead projector acetates, are used to display sites of similar flow rate, dissolved oxygen, weed cover values and depth.

11 Interpret the arrangement of sites in terms of flow rate, oxygen, weed cover and depth.

Exercise 52
Ordination of woodland ground flora

Apparatus

1m^2 quadrats; compasses; soil augers; rulers; pH meter; light meter; oven; crucibles; balance; random number tables; labels; polythene bags.

Preparation

1 The student should be familiar with the Domin scale of recording vegetation (see Exercise 6).

2 A woodland site is selected which will provide at least one marked environmental gradient, e.g. soil water, soil type, light, etc.

3 A species list is compiled and used to produce data sheets.

4 The ordination technique is programmed for a computer and set to accept the data in a convenient format.

Procedure

1 Students work in pairs, and each pair is allocated 5 numbers in sequence.

2 Using compasses and random number tables students locate 5 random sites in a woodland.

3 At each site the students place a 1 m^2 quadrat and,

(*a*) record the vegetation using the Domin scale.

(*b*) take a soil sample within the 1 m^2 quadrat using an auger. The depth of humus is recorded and the soil placed in a labelled polythene bag.

(*c*) record the quantity of light reaching the top of the vegetation.

4 Return to the laboratory.

5 Soil samples are examined, the pH tested and recorded and the soil

moisture estimated by drying to constant weight (see Chapman, 1976, for soil analysis techniques).

6 Compile data sets on the blackboard for each site, recording pH, light, humus depth, soil moisture.

7 Species data is tabulated for each site as,

Sites	*Species*			
	1	*2*	*3*	*4*
1	3	5	2	0
2	4	1	0	0
3	0	0	0	5
4	1	1	0	1
5	0	0	0	3

Table of Domin values.

8 Data are punched onto cards or other suitable computer input and entered in the computer as data for the ordination program.

9 If the program calculates only similarity values for the species pairings then a half matrix is erected and the ordination constructed as on p. 191.

10 If the computer is programmed to construct the half matrix and also to plot the ordination, continue at step 11.

11 Using tracing paper or overhead projector acetates, the sites are located and overlays produced of pH, humus depth, soil moisture and light values.

12 Interpret the site groupings in terms of habitat variables.

Discussion and conclusions

Ordination of freshwater sites based on the similarity of their species complement usefully summarizes and simplifies the information contained in the data. Usually freshwater organisms are examined in relation to their aquatic environment species by species and the results interpreted in terms of adaptations to flow rate, oxygen level etc. Here the *sites* are compared with each other on the basis of their species similarity.

This approach is very rewarding in that sites with apparently similar habitat characteristics can, on analysis, be found to possess a high degree of species dissimilarity. On returning to the field site clues are often evident to help explain these anomalies, such as small differences in micro-environment, eddies, effects of nearby aquatic vegetation or variation in flow with depth, producing rapid currents near the top but comparative calm near the bottom of a stream. This approach to sorting data is thus very useful when comparing sites for management purposes, whether they be aquatic, marine or terrestrial.

Where distinct habitat differences exist, woodland ground flora when ordinated tends to form clusters. However, since the method expresses the continuous variability in the composition of the ground flora, then the similarity between stands may result in an ordination diagram with a complete scatter of points rather than distinct clusters of stands. When any one axis is used this can result in a sequential arrangement of stands along the x axis. When this occurs, it usually reflects the fact that a single environmental trend dominates the community structure.

Suggestions for further work

1 As a laboratory study in ordination techniques any quantitative data from a range of sites may be used, including insect trap data or bird counts.

2 Many other ordination techniques exist, for instance those described in Kershaw (1973) and Shimwell (1971) which offer advances on the Bray and Curtis (1957) method. The most widely used of these is principal component analysis which efficiently displays the continuous variation of a community structure by constructing the first axis through the maximum variation present in the data. The second axis is drawn through the next highest variation. Several coefficients have been used to plot components but that offered by Orloci (1966) and Austin and Orloci (1966) appears to be useful for a wide range of problems.

$$\sum_{i=1}^{N} (x_{ij} - \bar{x}_i)(x_{ih} - \bar{x}_i)$$

where N = the number of species,
x_{ij} = the abundance of species i in sample j,
x_{ih} = the abundance of species i in sample h,
$\bar{x}_i$ = the mean abundance of species i.

The results of principal components analysis have been used successfully to postulate phytosociological relationships and community/habitat correlations. The method has the advantage of being robust enough to provide useful results even when only the most common species (75% of the total species complement) are recorded. This means that one can virtually ignore the rare species. See Kershaw (1973) for a detailed account of the methodology and examples.

3 Several different quantitative measures can be used to assess the quantity of vegetation on the ground (see Section 1). Ordinations of the same sites based on different measures of species abundance, e.g. density, cover, etc., could be performed to establish the effect of the vegetation measure on the resultant ordination. Qualitative data (i.e. presence and absence records) have also been used for ordinations.

4 Shimwell (1971) pp. 270–273 describes how species can be ordinated. In this case individual species are compared by estimating the similarity between species by calculating the C values on the basis of their different quantities in the samples.

Exercises 53 and 54

Classification of vegetation

Principles

One method of objectively extracting natural groupings in vegetation is by numerical classification based on some intrinsic property of the vegetation. Ecological classification is essentially a process which produces groups of similar stands by a process conceptually similar to the taxonomic classifications with which we are more familiar. Formation of these groups may be by various different strategies, most of which form hierarchical structures of the data. These hierarchical strategies may be termed either agglomerative, which take a set of individuals and progressively fuse them to form larger and larger groups, or divisive, which take the population as a whole and progressively divide this into smaller and smaller sub-groups. In these exercises a simple monothetic divisive strategy is described. This type of classification hierarchically divides a population of individuals (in this case quadrats) into groups, each defined by the presence or absence of an attribute (a plant species). At each division of the population the species selected for dividing the population is determined as that which maximizes the difference between individuals (quadrats) using a specific measure. For the method described below the measure used is the between-groups-sum-of-squares (ΔSS).

Data are collected as species presence and absence records in each quadrat. The quadrats then become the individuals of the population to be classified and the species complement of each quadrat its attributes. The ΔSS value for each species in turn is calculated from:

$$\Delta \mathrm{SS} = \frac{n_a \cdot n_b}{n_a + n_b} \sum_{k=1}^{p} (\bar{x}_{ak} - \bar{x}_{bk})^2$$

where n_a = number of individuals (quadrats) in sub-group a,
n_b = number of individuals (quadrats) in sub-group b,
p = total number of attributes (species) including the one being tested,
$\bar{x}_{ak}$ = proportion of samples in sub-group a containing the species k,
$\bar{x}_{bk}$ = proportion of samples in sub-group b containing species k.

Sub-group a contains the species being tested for discrimination; sub-group b does not contain the species being tested.

The method known as MONO is rapid and easy to program. Only small sets of data used for instruction should be attempted by pocket calculator. A population is investigated by dividing on each species in turn. The ΔSS is calculated for each species and the one which maximizes the statistic, i.e. gives the largest ΔSS value, is selected as the discriminant species and used to divide the population into two sub-groups. The procedure is then repeated on the sub-groups (i.e. attempting to sub-divide them further by using each species in turn as a potential discriminant) until the ΔSS value falls below 10, or the required degree of sub-division is reached. Since the sub-groups are defined by the presence or absence of discriminant species this type of classification has the advantage of supplying a simple dichotomous key to the vegetation groups.

In the laboratory study the data contained in the Atlas of British Flora or a county flora are used to examine natural groupings of vegetation on a large scale. A few quadrats (sites) are selected and the calculations carried out by the students with the aid of calculators. In contrast, the field study examines the small scale vegetation groupings in a woodland (although any natural or semi-natural vegetation is suitable). Here the larger data set should be analysed by computer. Results will allow the vegetation groups to be mapped and compared with habitat data.

Exercise 53
Separation of major vegetation groupings by numerical classification

Apparatus

A floral atlas, either national or county; data sheets; calculators; coloured felt tip pens.

Preparation

The objective of this exercise is to demonstrate the working of numerical classification. Because it is not practicable to analyse any but the smallest of data sets by hand it is difficult to provide a real ecological situation suitable for exploring the methodology. However in recent years the Botanical Society of the British Isles and several County Naturalists' Trusts have produced very detailed floral atlases. These may be used to select individual locations (sites) and record the presence or absence of individual species.

1 Make copies of the distribution map of as many species of plants as there are students. As this exercise is primarily to teach the methodology, the selection can be contrived to produce easily separable groupings, e.g. by selecting calcicoles/califuges.

2 Using random number tables identify 50 of the mapped squares to be used as sample sites and mark these squares on each map.

Procedure

1 Students are allocated one site and have to record the presence or absence of each of the selected species in this square.

2 Class results are tabulated on the blackboard as below:

	Species									
Site no.	A	B	C	D	E	F	G	H	I	J
1	+	0	+	+	0	+	+	0	0	0
2	0	0	0	0	+	0	0	+	+	0
3	+	+	+	+	0	0	+	0	0	0
4	+	+	0	+	0	+	+	0	0	+

3 Each student calculates the ΔSS value for a single species (see Principles section).

4 The species producing the highest ΔSS value is used to divide the population.

5 The two sub-groups are sub-divided using the same technique, i.e. producing 4 groups in all.

6 Construct a hierarchy from the results; see Fig. 53/54.1.

7 The original sites are then allocated to one of the four groups and marked on a map showing their location, using a different colour or symbol for each group.

8 If the BSBI Atlas of Flowering Plants has been used then overlays of climate, geology, etc. are available and should be used to investigate differences between the groups.

Exercise 54
Classification of woodland ground flora

Apparatus

Six 2 m bamboo canes or surveyors' poles; 50 m measuring tapes; 8 wooden stakes; 10 m^2 quadrats; data sheets; 6 × 100 m lengths of orange twine marked at 10 m intervals.

Preparation

1 Prior to the field visit the woodland site should be staked out to provide a skeleton grid. In the example described here the number of quadrats is determined by the number of students (taken as an example to be 20). A grid of 100 m × 50 m should be erected along some environmental gradient, e.g. slope or soil moisture gradient. The grid is marked at the corners by large poles and along the short sides by

wooden stakes at 10 m intervals. The orange twine is tied between the poles and between the wooden stakes as below.

1	2	3	4	5	6	7	8	9	10
11	12	13	14	15	16	17	18	19	20
21	22	23	24	25	26	27	28	29	30
31	32	33	34	35	36	37	38	39	40
41	42	43	44	45	46	47	48	49	50

Direction of environmental gradient ⟶

Diagram showing the layout of skeleton grid (and the sequential numbering of grid squares) for vegetation survey.

2 A species list is compiled and each species given a code number. New species found during the field exercise are added to the list.

3 Prepare data sheets and outline maps of the grid.

4 Program the MONO classification.

Procedure

1st afternoon

1 Students work in pairs.

2 The 1 m^2 quadrats are placed (at random) within each of the 10 × 10 m grid squares along the top row. Students record the presence of plants growing in their quadrat.

3 Collect environmental data for each quadrat, e.g. soil pH and humidity, humus depth, light, and canopy composition.

4 When each pair has finished recording for their quadrat they move down to the next row, i.e. if a student pair starts with grid square no. 1 they will move to 11, 21, 31, 41 and students starting at 6 will move to 16, 26, 36, 46.

5 Data sheets are standardized and 'new species' given a code number.

2nd afternoon

1 The data are arranged for analysis as below, ignoring all species with less than 4 quadrat records (or 5% if over 100 quadrats are used). This will reduce computer storage and time and, in all but the most critical studies, produce satisfactory results.

Quadrat number	*Attributes (spp.) present*
1	1, 6, 7, 8, 10, 11, 12, 22, 26, 27, 28, 29
2	5, 8, 15, 16, 17, 18, 19, 21, 24, 25
3	1, 2, 6, 7, 8, 10, 11, 12, 22, 26, 27, 29, 31
4	5, 6, 8, 14, 15, 16, 17, 18, 20, 21, 24, 29
5	5, 8, 9, 14, 15, 16, 17, 18, 19, 20, 21, 24, 25

2 Code the data into a suitable computer input, whether on card, tape or disc. Students are arranged to take a fair share of this work.

3 Input data into the classificatory procedure MONO which should be filed as a computer package. For this exercise it suffices to divide the population into 8 groups.

4 Construct a hierarchy from the results; see Fig. 53/54.1.

5 Draw maps of the study area (i.e. the grid) and mark on the sites allocated to the vegetation groups at the 2, 3, 4, 5, 6, 7 and 8-group levels as identified by the program. This task may be divided between groups of students to save time.

6 Examine the maps and identify the point at which the contiguity between adjacent quadrats starts to break down into a mosaic of individual quadrats. It is rarely worth dividing a population below this level since interpretation becomes increasingly difficult.

7 Interpret vegetation maps by comparison with environmental data. Formulate hypotheses to explain the distribution of ground flora. How could these be tested?

Discussion and conclusions

Laboratory studies attempting to employ classificatory techniques are of real value in teaching the concepts involved and can also provide useful structuring of ecological data that exists in distribution atlases. The BTO Atlas of Breeding Birds in Britain and Ireland used for the χ^2 classification in Exercise 45 could also be used as a data set for classification.

The classification of woodland ground flora can be used to demonstrate the potential of the classificatory approach to vegetation survey. The results should be readily reproducible since each stage is objective, the final groups are distinct and are directly amenable to both vegetation mapping and comparison with habitat variables.

Scale is an important consideration, for classification procedures may work at a very large scale separating major biomes, such as woodland from heathland, down to the separation of phytosociological units present in chalk grassland. It is easy to demonstrate this feature by running two exercises concurrently, one taking random samples within a km^2 and the other within a 100 m × 100 m area within the larger area. The data is classified to the same group level (e.g. 8 groups) and comparisons made. For some survey work it would be profitable to undertake a multistage classification such as this. However the scale used is usually arrived at by reference to the objectives of the study.

The study objectives also enable one to decide upon the most suitable numerical technique to use to analyse or describe a particular vegetation type. Goldsmith and Harrison (1976) offer a key to numerical techniques based on study objectives, which enables the user to establish quickly the most appropriate technique to solve a particular problem of analysis. Recent advances in classificatory techniques appear in journals such as *Vegetatio* and the *Journal of Ecology*. These journals contain papers by workers who have investigated the role of different coefficients used to divide the population, the effects of quantitative vs. qualitative data, combinations of both the ordination and classificatory approaches used to structure data, the ways of comparing one hierarchy with another and techniques of comparing vegetation groups with habitat data.

Further investigations

1 The use of classification for mapping is well proven and when a monothetic divisive strategy such as MONO is employed it has the further advantage of providing a dichotomous key to vegetation communities, e.g. if the hierarchy shown in Figure 53/54.1 resulted from a classification, then the following key could be constructed:

1. *Galium aparine*	present	2
	absent	5
2. *Mercurialis perennis*	present	3
	absent	4
3. *Cercia lutetiana*	present	vegetation group ①
	absent	vegetation group ②
4. *Rumex acetosa*	present	vegetation group ③
	absent	vegetation group ④
5. *Anemone nemorosa*	present	6
	absent	7
6. *Oxalis acetosella*	present	vegetation group ⑤
	absent	vegetation group ⑥
7. *Dryopteris dilitata*	present	vegetation group ⑦
	absent	8
8. *Pteridium aquilinum*	present	vegetation group ⑧
	absent	vegetation group ⑨

Keys such as this can then be used in the field to allocate vegetation to one or other of the classified groups. In this way large areas of vegetation

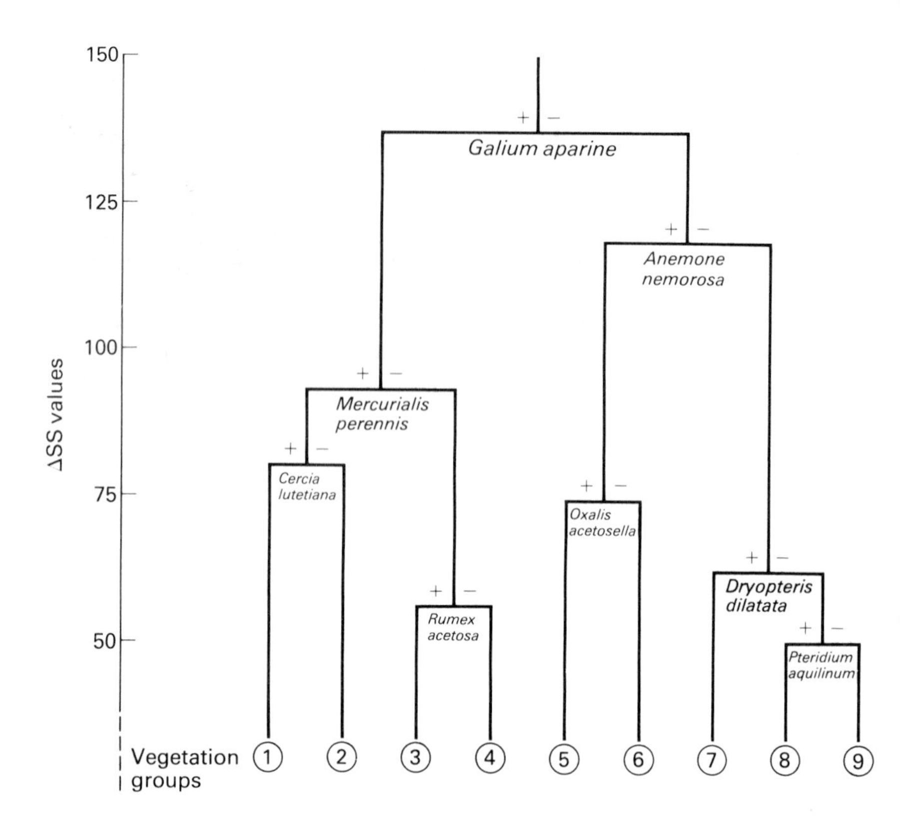

Fig. 53/54.1 An example of a hierarchy resulting from a MONO classification of woodland ground flora.

can be mapped. To obtain the most objective initial survey, random or stratified random samples should be collected throughout the area to be mapped for the initial classification.

2 A very obvious sequel to vegetation classification and mapping is the detailed recording of habitat variables from the same area. Accurate maps of pH, moisture and light gradients enable the formulation of hypotheses pertaining to the distribution of the plant communities identified by classification. Further field work relating performance of individual species to habitat factors and employing statistical regression might lead to the establishment of causal relationships. In many situations, comparison of maps showing the distribution of pH, soil nutrients or soil moisture and those showing the distribution of vegetation groups will provide sufficient stimulus for much speculation and a possible starting point for long term studies.

3 It is possible to use many different coefficients to subdivide a population; one of the first and most widely used is χ^2 as used by

Williams and Lambert (1959) in their association analysis; another popular coefficient is the information statistic of Lance and Williams (1968) in which the species producing the greatest drop in information (ΔI) on sub-division is selected as the discriminant species at each stage. Using the same data set it is possible to obtain subtly different classifications using these different coefficients. A comparison of the different hierarchies provides an insight into the characteristics of each.

4 Any monothetic divisive strategy is likely to provide some misclassifications since the allocation of an individual quadrat to one group or another depends solely on the presence or absence of the discriminant species. It is obvious that a quadrat having a species complement characteristic of a group, but not the discriminant species, will be misclassified. One method of overcoming this problem, which takes into account the total species complement of each quadrat, is described in Lambert *et al.* (1973). The method, MONIT, described in this paper is closely related to the method MONO described in this section and provides a useful comparison when the two are used on the same data set.

5 Since both the sampling and analytical stages of a numerical classification are objective it is a useful exercise to get students to spend an afternoon subjectively mapping the vegetation of an area and comparing the results with those from a classification. In some cases it is difficult to argue in favour of the numerical approach if the communities are obvious by eye, as the time taken to collect and analyse the data for the numerical classification is much greater than for a descriptive survey. Another point worthy of consideration is the delineation of boundaries in vegetation mapping. Objective sampling in an area suffers from the drawback that it is not sensitive to sharp changes in the vegetation and cannot accurately reflect the true shape of community boundaries when a grid sampling technique is used. As yet no one has provided a workable objective method for mapping the true boundaries of vegetation communities and this problem could provide a number of interesting projects.

6 Although it has been employed primarily for vegetation survey, classification has many other applications in ecology. Detecting associations between animals or the grouping of habitat variables have in the past employed classificatory techniques and they are increasingly used in the social sciences to classify people into groups possessing common attributes. However as a numerical tool for ecologists, geographers, social scientists, etc. they are capable of even wider applications than at present. In fact there are many situations where a 'population' requires sorting into groups by the presence or absence of their attributes that could be investigated by classification.

Exercises 55 and 56

The ecology of plant-herbivore relationships

Principles

Although it is virtually unknown for a plant species to support no herbivorous insects, there are some species which are noticeably 'undergrazed'. Many plants have obvious physical adaptations which seem to limit feeding by insects (spines, thick cuticle, hairs etc.) while others possess one or more chemicals with no obvious metabolic role in the plant but which repel potential herbivorous insects. Such 'odd substances' or secondary plant substances (see Exercise 29) do not appear to be universal, however, and it has been suggested (Cates and Orians, 1975) that their occurrence, and therefore plant palatability, is related to the successional status of the plant (early successional pioneer species, climax species etc.). This hypothesis stems from the observation that early successional plant communities tend to be comparatively shortlived. Therefore one could suggest that the plants have evolved for rapid growth and development and for maximum reproductive potential; all available resources are devoted to reproduction. Climax communities, on the other hand, are more stable; their plant species may persist for longer and individual plants are longer-lived. Increased longevity provides the possibility of reproducing in more than one season, so there is less intense selection to maximize reproductive potential in any one year. Some resources may therefore be devoted to other uses, one of which could be the provision of chemical defence mechanisms. The laboratory exercise tests this hypothesis by exposing plant species of differing successional status to a general herbivore (snail or slug) and determining palatability.

If the above hypothesis is valid, it does not follow that all late-successional or climax species, such as many trees, are equally resistant to herbivores. An examination of the number of herbivorous insects associated with various tree species shows that there is, in fact, great variation in the number of insects which feed on them (Southwood, 1961, 1972). One might expect the number of herbivorous species associated with a tree to be related to the time the tree has been in the country and, perhaps more importantly in evolutionary terms, its general abundance or scarcity throughout the period of its occurrence. This suggestion is based on the thesis that the insect/plant relationship is a continuously evolving one, and the greater the number of occasions that insects have been exposed to a particular repellent or toxic chemical the more likely are they to have evolved a mechanism for dealing with it (Southwood, 1972). Indeed, in some cases, insects appear not only to have overcome the substance's harmful properties but now use it as a key chemical by which they identify their host plant (see Exercise 29). Southwood (1961) quantified the abundance of some tree species in Britain during the Quaternary era by referring to Godwin's (1956) compilation of all records of plant remains during this period. He found that the number of herbivorous insect species recorded from each tree species (derived from the literature) was related to the tree's 'recent' abundance. This appeared to support the hypothesis. Exercise 56 tests the same hypothesis by collecting insects from various trees in a standard way and relating the number captured to the trees' recent geological abundance (Godwin's data).

Exercise 55
Successional status and the palatability of plants to general herbivores

Apparatus

Supplies of plant leaf material, fresh or frozen in air-less polythene bags. Examples of species within three main divisions of successional status are: early successional: clover (*Trifolium repens*), dandelion (*Taraxacum officinale*), hogweed (*Heracleum sphondylium*), plantain (*Plantago* spp.), dock (*Rumex* spp.); mid successional: bramble (*Rubus fruticosus*), willow (*Salix* spp.), birch (*Betula* spp.), alder (*Alnus glutinosa*), hawthorn (*Crataegus monogyna*); late successional: oak (*Quercus robur* or *petraea*), ash (*Fraxinus excelsior*), beech (*Fagus sylvatica*), sycamore (*Acer pseudoplatanus*). Provide as many 1 cm diameter cork borers as possible; graph paper (mm^2). For each student pair: four plastic boxes (*c.* 30 × 20 × 10 cm) with lids and moist soil to a depth of 2 cm; 4 snails or slugs (the whole class should use the same species); the molluscs may be kept in aquaria and fed lettuce, dandelion etc. for a long period before the practical, if necessary. Deprive them of food for 12 hours before the exercise.

Preparation

1 Provide each student pair with 4 boxes each containing one snail or slug. Provide each pair with leaf material representing 8 species: 4 from each of 2 successional classes. It is not vital that every species is compared with every other, nor that there is a similar number of plant species in each successional group.

2 A trial exercise to determine the time for which the herbivores should be left in the boxes is advisable, as their feeding rate cannot be

predicted (see Procedure). If too many discs (see Procedure) are completely consumed even in the shortest convenient time interval, increase the disc size or number.

Procedure

1 For the first of the four boxes, cut eight leaf discs (two from each of four of the plant species provided). Examine the discs carefully, using a binocular microscope if necessary, and make notes of each species' characteristics in disc form.

2 If the leaves of some plant species are not large enough to cut a 1 cm diameter disc, cut a portion as near to 1 cm as possible, trace round it on graph paper and count the squares to determine its area. The area of a disc of 1 cm diameter is 79 mm^2.

3 Place the discs in the container, add a single snail or slug and replace the lid.

4 Repeat 1–3 with the same plant species in the second box.

5 Repeat 1–3 for the other four plant species in each of the remaining two boxes.

6 Return to the boxes the next day, the organizer having removed the snails or slugs by this time if the trial exercise revealed a significant consumption of at least some discs well within the time between the two student periods.

7 Remove and identify each disc or disc-fragment in turn and measure its area, if some has been consumed, by drawing round it on graph paper and counting the squares.

8 Enter the results (in mm^2 consumed) in a blackboard tabulation.

9 When the results from all replicates have been tabulated, add up the data in each category and prepare a second table as follows:

	Total mm^2 consumed		
	EARLY	*MID*	*LATE*
EARLY vs. MID			
No. replicates			
Mean mm^2/replicate			
MID vs. LATE			
No. replicates			
Mean mm^2/replicate			
LATE vs. EARLY			
No. replicates			
Mean mm^2/replicate			
TOTAL mm^2 consumed			
No. replicates			
Mean mm^2/replicate			

Exercise 56
The number of herbivorous insect species associated with various trees

Apparatus

Per pair of students: a beating tray plus beating stick and 2 pooters (see Southwood 1971), an inverted umbrella may be substituted for the beating tray; 20–30 corked specimen tubes; sticky labels or scraps of paper; pencils.

Preparation

1 Find several specimens of as many of the trees in the list below (from Godwin, 1956) as possible. Mixed woodland, parkland and large gardens should together provide most of them. Locate enough of each species so that each student pair can sample from most species without any tree being visited more than once. If time is limited, try to include species from each end of the list.

2 The list below summarizes Godwin's (1956) data:

	Number of records in Godwin (1956)
Oak (*Quercus robur* and *Q. petraea*)	197
Birch (*Betula* spp.)	182
Hazel (*Corylus avellana*)	136
Willow (*Salix* spp.)	134
Alder (*Alnus glutinosa*)	87
Hawthorn (*Crataegus* spp.)	67
Ash (*Fraxinus excelsior*)	59
Pine (*Pinus sylvestris*)	54
Holly (*Ilex aquifolium*)	44
Yew (*Taxus baccata*)	42
Sloe (*Prunus spinosa*)	30
Poplars (*Populus* spp.)	30
Elm (*Ulmus* spp.)	30
Beech (*Fagus sylvatica*)	27
Common maple (*Acer campestre*)	18
Hornbeam (*Carpinus betulus*)	17
Juniper (*Juniperus communis*)	17
Spruce (*Picea abies*)	15
Lime (*Tilia* spp.)	14
Mountain ash (*Sorbus aucuparia*)	13

continued

continued

Fir (*Abies* spp.)	10
Sweet chestnut (*Castanea sativa*)	10
Apple (*Malus* spp.)	7
Walnut (*Juglans regia*)	3
Holm oak (*Quercus ilex*)	2
Larch (*Larix decidua*)	1
Sycamore (*Acer pseudoplatanus*)	1
Horse-chestnut (*Aesculus hippocastanum*)	11
Acacia (*Robinia pseudacacia*)	0
Plane (*Platanus orientalis*)	0

Procedure

1 Each student pair should visit as many tree species as possible. For each species in turn and with the pooter tube already in the mouth, hold the tray beneath a branch (without knocking the branch) and beat the branch firmly from above about three times. Fragments of leaf and twig material will fall onto the tray along with insects.

2 Immediately place the tray on the ground. Both students should pooter up all the insects from the tray as quickly as possible concentrating on the most active individuals first.

3 Cork the pooter collecting tube and replace it with an empty one. Label the tube with the tree species from which its contents came.

4 Repeat 1–3 on the same tree if another low branch is available, or another tree of the same species. Compare the pair of tubes from the second catch visually with those from the first and try to decide whether a new herbivorous species was captured during the second sample. If difficulty is experienced classifying the insects according to feeding habits (this needs entomological experience) decide simply whether a new species was captured.

5 Repeat 1–3 for the same tree species until no new insect species (herbivorous or all types) has been captured for three successive samples. If assistance is available, count the total number of presumed herbivorous insect species captured. Alternatively, when all tree species have been sampled take the catch back to the laboratory where reference books or further assistance should be available to classify the catch according to diet.

6 Repeat the above procedure for as many tree species as possible.

7 In the laboratory, enter on the blackboard the number of herbivorous insects captured on each tree species. The organizer should then calculate the mean for each tree species visited.

8 Each student should plot the above mean values against the index of the tree species' evolutionary abundance (Godwin's data). Calculate the regression equation (Parker, 1979) if required.

Discussion and conclusions

There has been considerable controversy in recent years concerning the attributes of plants which can affect their attractiveness and palatability to insects. Their nutritional status and 'odd substance' content are both involved and their relative importance varies with the plant and insect species (Fraenkel, 1969; Thorsteinson, 1960). The results with the snails or slugs in Exercise 55 should show that to a *general* herbivore (i.e. one which has not evolved to respond to a particular nutritional balance or odd substance), late successional species are less palatable than early ones and it is tempting to relate this to the supposed wider range and increased quantities of secondary plant substances in this group.

Predictions should also be confirmed in Exercise 56, but the most interesting species in the regression are those which fall considerably below the line. These appear to have defence mechanisms which insects find particularly difficult to overcome. There could be other reasons for the scatter of points, however, especially the restricted nature of the sampling in Exercise 56. For instance, when this exercise was carried out recently, the result for birch fell well below the regression line. As the samples were taken at the end of an exceptionally hot, dry summer, the low insect catch from birch could have been a reflection of the water stress suffered by this shallow-rooted species; its leaves were yellowish-green and may have been an unusually poor food source in that year.

Considering that Southwood's literature survey revealed figures as high as 284 herbivores for oak, and that the methods in Exercise 56 rarely produce more than 30–40, it is interesting that the expected relationship can be demonstrated so easily.

Further investigations

1 A field exercise concerning palatability could involve collections of insects by sweeping or beating from plants in all successional stages. The insects collected should be classified, with the aid of identification guides or experienced entomologists, as species specific to the plant on which they were captured, or general feeders. The total number of species in the latter category could then be related to the hosts' successional status.

2 There is some evidence that the peak of insect grazing activity on oak is related to the concentration of tannins and protein in the leaves (Feeny, 1970), with the result that most caterpillar feeding occurs when the leaves are young and low in tannin. By freezing weekly samples of oak leaves from bud-burst to senescence, one could modify the leaf disc experiment and measure the palatability of different-aged oak leaves.

3 Another series of leaf discs could be prepared, preferably all from species in the same successional class, in which there is a range of cuticle thickness or hairiness. Palatability could be related to this.

4 The tree species found to have many fewer insect species on them than predicted by the regression line in Exercise 56 could be offered as leaf discs to the snails or slugs and their palatability compared with that of those species which lie well above the regression line. This could provide evidence that their position above or below the line really is related to their chemical resistance to grazing rather than to a sampling artefact.

5 If the successional status theory of Cates and Orians (1975) is correct, a comparison of two con-generic species of plant, one an annual and the other a perennial or biennial, would be of interest. The snail or slug grazing technique could be used and a suitable con-generic pair of plants would be: *Urtica urens* (small nettle—annual) and *Urtica dioica* (stinging nettle—perennial).

References

Agnew, A. D. Q. (1961). The ecology of *Juncus effusus* L. in North Wales. *J. Ecol.*, **49**, 83–102.

Allen, J. A. (1972). Evidence for stabilizing and apostatic selection by wild blackbirds. *Nature, Lond.*, **237**, 348–349.

Allen, J. A. (1974). Further evidence for apostatic selection by wild passerine birds: training experiments. *Heredity*, **33**, 361–372.

Allen, J. A. and Clarke, B. (1968). Evidence for apostatic selection by wild passerines. *Nature, Lond.*, **220**, 501–502.

Anderson, D. J. (1967). Studies on structure in plant communities. III Data on pattern in colonizing species. *J. Ecol.*, **55**, 397–404.

Austin, M. P. and Orloci, L. (1966). Geometric models in ecology. II An evaluation of some ordination techniques. *J. Ecol.*, **54**, 217–227.

Bailey, N. T. J. (1959). *Statistical Methods in Biology*. English Universities Press, London.

Ballentine, W. J. (1961). A biologically-defined exposure scale for the comparative description of rocky shores. *Field Studies*, **1**, 1–19.

Bantock, C. R. and Harvey, P. H. (1974). Colour polymorphism and selective predation experiments. *J. Biol. Educ.*, **8**, 323–329.

Barclay-Estrup, P. and Gimingham, C. H. (1969). The description and interpretation of cyclical processes in a heath community. 1. Vegetational change in relation to the *Calluna* cycle. *J. Ecol.*, **57**, 737–758.

Birch, L. C. (1957). The meanings of competition. *Am. Nat.*, **91**, 5–18.

Blackman, G. E. (1935). A study by statistical methods of the distribution of species in grassland association. *Ann. Bot. Lond.*, **49**, 749–778.

Bradley, R. D., Christie, J. M. and Johnston, D. R. (1966). Forest Management Tables. *Bookl. For. Commn.*, **16**.

Bray, R. J. and Curtis, J. T. (1957). An ordination of the upland forest communities of Southern Wisconsin. *Ecol. Monogr.*, **27**, 325–349.

Brower, J. U. Z. (1960). Experimental studies of mimicry. 4. The reactions of starlings to different proportions of models and mimics. *Am. Nat.*, **94**, 271–282.

Butler, E. J. and Jones, S. G. (1949). *Plant Pathology*, MacMillan.

Cairns, J., Albaugh, D. W., Busey, F. and Chaney, D. (1968). The sequential comparison index—a simplified method for non-biologists to estimate relative differences in biological diversity in stream pollution. *J. Wat. Pollut. Control Fed.*, **40**, 1607–1613.

Cates, R. G. and Orions, G. H. (1975). Successional status and the palatability of plants to generalized herbivores. *Ecology*, **56**, 410–418.

Caughley, G. (1966). Mortality patterns in mammals. *Ecology*, **47**, 906–917.

Chapman, S. B. (Ed.) (1976). *Methods in Plant Ecology.* Blackwell, Oxford.

Chinery, M. (1973). *A Field Guide to the Insects of Britain and Northern Europe.* Collins, London.

Clarke, B. (1962). Balanced polymorphism and the diversity of sympatric species. *Taxonomy and Geography.* Edited by Nichols, D. The Systematics Association, London.

Clark, P. J. and Evans, F. C. (1954). Distance to nearest neighbour as a measure of spatial relationships in population. *Ecology.*, **35**, 445–453.

Clapham, A. R., Tutin, T. G. and Warburg, E. F. (1962). *Flora of the British Isles.* 2nd Ed. C.U.P., Cambridge.

Cloudsley-Thompson, J. L. and Sankey, J. (1961). *Land Invertebrates.* Methuen, London.

Connell, J. H. (1961a). The effects of competition, predation by *Thais lapillus* and other factors on natural populations of the barnacle, *Balanus balanoides. Ecol. Monogr.*, **31**, 61–104.

Connell, J. H. (1961b). The influence of interspecific competition and other factors on the distribution of the barnacle *Chthalamus stellatus. Ecology*, **42**, 710–723.

Croze, H. (1970). Searching image in carrion crows. *Z. Tierpsychol.*, supplement **5**, 1–86.

Dean, G. J. W. (1973). Distribution of aphids in spring cereals. *J. appl. Ecol.*, **10**, 447–462.

Deevey, E. S. (1947). Life tables for natural populations of animals. *Quart. Rev. Biol.*, **22**, 283–314.

Dempster, J. P. (1971). The population ecology of the cinnabar moth, *Tyria jacobaeae* L. (Lepidoptera, Arctiidae). *Oecologia*, **7**, 26–67.

Dempster, J. P. (1975). *Animal Population Ecology.* Academic Press, London.

Dempster, J. P. and Coaker, T. H. (1974). Diversification of crop ecosystems as a means of controlling pests. *Biology in Pest and Disease Control. Symposia of the British Ecological Society No. 13.* Edited by Jones, D. P. and Solomon, M. E. Blackwell, Oxford.

Dixon, A. F. G. (1966). The effect of population density and nutritive status of the host on the summer reproductive activity of the sycamore aphid, *Drepanosiphum platanoides* (Schr.). *J. Anim. Ecol.*, **35**, 105–112.

DIXON, A. F. G. (1969). Population dynamics of the sycamore aphid *Drepanosiphum platanoides* (Schr.) (Hemiptera: Aphididae): Migratory and trivial flight activity. *J. Anim. Ecol.*, **38**, 585–606.

DIXON, A. F. G. and LOGAN, M. (1973). Leaf size and availability of space to the sycamore aphid *Drepanosiphum platanoides*. *Oikos*, **24**, 58–63.

DIXON, A. F. G. and MCKAY, S. (1970). Aggregation in the sycamore aphid *Drepanosiphum platanoides* (Schr.) (Hemiptera: Aphididae) and its relevance to the regulation of population growth. *J. Anim. Ecol.*, **39**, 439–454.

DIXON, A. F. G. and WRATTEN, S. D. (1971). Laboratory studies on aggregation, size and fecundity in the black bean aphid, *Aphis fabae* Scop. *Bull. ent. Res.*, **61**, 97–111.

DOWDESWELL, W. H. (1959). *Practical Animal Ecology*. Methuen, London.

EBLING, F. J., KITCHING, J. A., MUNTZ, L. and TAYLOR, C. M. (1964). The ecology of Lough Ine XIII. Experimental observations of the destruction of *Mytilus edulis* and *Nucella lapillus* by crabs. *J. Anim. Ecol.*, **33**, 73–82.

EDMUNDS, M. (1974). *Defence in animals*. Longman, London.

ELLIOTT, J. M. (1977). *Some methods for the statistical analysis of samples of benthic invertebrates*. Freshwater Biological Association.

ERRINGTON, J. C. (1973). The effect of regular and random distributions on the analysis of pattern. *J. Ecol.*, **61**, 99–105.

FEENY, P. P. (1970). Seasonal changes in oak leaf tannins and nutrients as a cause of spring feeding by winter moth caterpillars. *Ecology*, **51**, 565–581.

FORD, E. B. (1964). *Ecological genetics*. Methuen, London.

FRAENKEL, G. (1969). Evaluation of our thoughts on secondary plant substances. *Ent. exp. & appl.*, **12**, 473–486.

FREELAND, P. W. (1970). The productivity of the Scots Heather (*Calluna vulgaris*) in the *Calluna-Ulex minor* complex of Ashdown Forest, Sussex. *J. Biol. Educ.*, **4**, 297–304.

GARB, S. (1961). Differential growth-inhibitors produced by plants. *Bot. Rev.*, **27**, 422–443.

GAUSE, G. F. (1934). *The Struggle for Existence*. Hafner, New York (Reprinted 1964).

GIMINGHAM, C. H. (1975). *An Introduction to Heathland Ecology*. Oliver and Boyd, Edinburgh.

GODWIN, H. (1956). *The History of the British Flora*. Cambridge University Press, Cambridge.

GOLDSMITH, F. B. and HARRISON, C. M. (1976). Description and Analysis of Vegetation. In CHAPMAN, S. B. (Ed.) *Methods in Plant Ecology*. Blackwell, Oxford.

GREENSLADE, P. J. M. (1964). The distribution, dispersal and size of a population of *Nebria brevicollis* (F.) with comparative studies on three other Carabidae. *J. Anim. Ecol.*, **33**, 311–333.

GREIG-SMITH, P. (1952). The use of random and contiguous quadrats in the study of the structure of plant communities. *Ann. Bot., Lond., N.S.* **16**, 293–316.

GREIG-SMITH, P. (1961). Data on pattern within plant communities. 1. The analysis of pattern. *J. Ecol.*, **49**, 695–702.

GREIG-SMITH, P. (1964). *Quantitative Plant Ecology.* 2nd Ed. Butterworths, London.

GRIME, J. P. (1974). Vegetation classification by reference to strategies. *Nature*, **250**, 26–31.

HARPER, J. L. and MCNAUGHTON, J. H. (1962). The comparative biology of closely related species living in the same area. VII Interference between individuals in pure and mixed populations of *Papaver* species. *New Phytol.*, **61**, 175–188.

HOLLING, C. S. (1959). Some characteristics of simple types of predation and parasitism. *Can. Ent.*, **91**, 385–398.

HOPKINS, B. (1957). Pattern in the plant community. *J. Ecol.*, **45**, 451–463.

HOUSE, H. L. (1969). Effects of different proportions of nutrients on insects. *Ent. exp. appl.*, **12**, 651–699.

JACKSON, R. M. and RAW, F. (1966). *Life in the Soil.* Edward Arnold, London.

JOLLY, G. M. (1965). Explicit estimates from capture-recapture data with both dead and immigration-stochastic model. *Biometrika*, **52**, 225–247.

KENNEDY, J. S. (1961). A turning point in the study of insect migration. *Nature, Lond.*, **189**, 785–791.

KERNEY, M. P. and CAMERON, R. A. D. (1979). *A Field Guide to the Land Mollusca of North-west Europe.* Collins, London.

KERSHAW, K. A. (1957). The use of cover and frequency in the detection of pattern in plant communities. *Ecology*, **38**, 291–299.

KERSHAW, K. A. (1961). Association and co-variance analysis of plant communities. *J. Ecol.*, **49**, 643–654.

KERSHAW, K. A. (1973). *Quantitative and Dynamic Plant Ecology*, 2nd Edition. Arnold, London.

KEVAN, D. K. MCE. (1962). *Soil Animals.* H. F. Witherby and Co., London.

KITCHING, J. A., MUNTZ, L. and EBLING, F. J. (1966). The ecology of Lough Ine. XV. The ecological significance of shell and body forms in *Nucella. J. Anim. Ecol.*, **35**, 113–126.

LAMBERT, J. M., MEACOCK, S. E., BARRS, J. and SMARTT, P. F. M. (1973). Axor and Monit: Two new polythetic-divisive strategies for hierarchical classification. *Taxon*, **22**, 173–176.

LANCE, G. N. and WILLIAMS, W. T. (1968). Note on a new Information Statistic Classificatory Program. *Comput. J.*, **11**, 195.

LEVIN, S. A. (1976). Spatial patterning and the structure of ecological communities. *Lectures on Mathematics in the Life Sciences*, **8**, 1–36.

LEWIS, T. (1965). The effects of an artificial windbreak on the aerial distribution of flying insects. *Ann. appl. Biol.*, **55**, 503–512.

LEWIS, T. and TAYLOR, L. R. (1967). *Introduction to Experimental Ecology*. Academic Press, London.

LINCOLN, F. C. (1930). Calculating waterfowl abundance on the basis of banding returns. *U.S.D.A. Circ.*, **118**, 1–4.

MALTHUS, T. R. (1798). *An Essay on the Principle of Population.* Reprinted by Macmillan, New York.

MCINTOSH, R. P. (1967). An index of diversity and the relation of certain concepts to diversity. *Ecology*, **48**, 392–404.

MERTZ, D. B. (1972). The *Tribolium* model and the mathematics of population growth. *Ann. Rev. Ecol. Syst.*, **3**, 51–78.

MILLER, C. A. (1963). The Spruce budworm. In *The Dynamics of Epidemic Spruce Budworm Populations*, R. F. MORRIS (Ed.). *Mem. Entomol. Soc. Canada* No. 31.

MULLER, C. H. (1966). The role of chemical inhibition (allelopathy) in vegetational composition. *Bull. Torrey Bot. Club*, **93**, 332–351.

MULLER, C. H., HANWATT, R. B. and MCPHERSON, J. K. (1968). Allelopathic control of herb growth in the fire cycle of California chaparral. *Bull. Torrey Bot. Club*, **95**, 225–231.

MULLER, C. H. (1970). Phytotoxins as plant habitat variables. *Recent Advances Phytochem.*, **3**, 105–121.

MURDOCH, W. W. (1973). The functional response of predators. *J. appl. Ecol.*, **10**, 335–342.

NICHOLSON, A. J. (1933). The balance of animal populations. *J. Anim. Ecol.*, **2**, 132–178.

ORLOCI, L. (1966). Geometric models in ecology. 1. The theory and application of some ordination methods. *J. Ecol.*, **54**, 193–215.

OVERLAND, L. (1966). The role of allelopathic substances in the 'smother crop' barley. *Amer. J. Bot.*, **53**, 423–432.

PARKER, R. E. (1979). *Introductory Statistics for Biology*, 2nd edition. Studies in Biology No. 43, Edward Arnold, London.

PARR, M. J., GASKELL, T. J. and GEORGE, B. J. (1968). Capture-recapture methods of estimating animal numbers. *J. Biol. Educ.*, **2**, 95–117.

PERRING, F. H. (Ed.) (1974). *The Biology of Bracken.* Academic Press, New York.

PICKERING, S. (1917). The effect of one plant on another. *Ann. Bot.*, **31**, 181–187.

PIELOU, E. C. (1964). *Population and Community Ecology: Principles and Methods.* Gordon and Breach, New York.

POLLARD, E., HOOPER, M. D. and MOORE, N. W. (1974). *Hedges.* Collins, London.

PROEBSTING, E. L. (1950). A case history of a 'peach repellent' situation. *Proc. Amer. Soc. Hort. Sci.*, **56**, 46–48.

RAMSBOTTOM, J. (1953). *Mushrooms and Toadstools.* Collins, London.

RANWELL, D. (1974). *Saltmarsh and Sand Dunes.* Chapman and Hall, London.

RAUNKIAER, C. (1934). *The Life Forms of Plants and Statistical Plant Geography.* Translated by Carter, Fausboll and Tansley. Oxford University Press, Oxford.

RICE, E. L. (1974). *Allelopathy.* Academic Press, London.

ROGERS, D. J. (1972). Random search and insect population models. *J. Anim. Ecol.*, **41**, 369–383.

RYTHER, J. H. (1954). Inhibitory effects of phytoplankton upon the feeding of *Daphnia magna* with reference to growth, reproduction and survival. *Ecology*, **35**, 522–533.

SCHOONHOVEN, L. M. (1972). Plant recognition by lepidopterous larvae. *Insect/Plant Relationships. Symposia of the Royal Entomological Society of London. No. 6.* Edited by VAN EMDEN, H. F. Blackwell, Oxford. pp. 87–99.

SCOURFIELD, D. J. and HARDING, J. P. (1966). *A Key to the British Species of Freshwater Cladocera.* Freshwater Biological Association. Scientific Publication No. 5.

SHARROCK, J. T. R. (1976). *The Atlas of Breeding Birds in Britain and Ireland.* Poyser, Berkhampsted.

SHEPPARD, P. M. (1951). Fluctuations in the selective value of certain phenotypes in the polymorphic land snail, *Cepaea nemoralis* (L.). *Heredity*, **5**, 125–134.

SHIMWELL, D. W. (1971). *Description and Classification of Vegetation.* Sidgwick and Jackson, London.

SLOBODKIN, L. B. (1962). *Growth and Regulation of Animal Populations.* Holt Rinehart and Winston, New York.

SOLOMOM, M. E. (1976). *Population Dynamics.* Studies in Biology No. 18. 2nd Edition. Edward Arnold, London.

SOUTH, R. (1961). *The Moths of the British Isles.* Warne, London.

SOUTHWOOD, T. R. E. (1961). The number of species of insects associated with various trees. *J. Anim. Ecol.*, **30**, 1–8.

SOUTHWOOD, T. R. E. (1962). Migration of terrestrial arthropods in relation to habitat. *Biol. Rev.*, **37**, 171–214.

SOUTHWOOD, T. R. E. (1972). The insect/plant relationship—an evolutionary perspective. In *Insect/Plant Relationships, Symposia of the Royal Entomological Society of London. No. 6.* Edited by VAN EMDEN, H. F., Blackwell, Oxford. pp. 3–30.

SOUTHWOOD, T. R. E. (1971). *Ecological Methods.* Chapman and Hall, London.

SVEDBURG, T. (1922). Ettibidrag till de statiska metodernas användning inom vöxtbiologien. *Svensk bot. Tidskr*, **16**, 1–8.

TABER, R. D. (1971). Criteria of sex and age. In HILES, R. G. (ed) *Wildlife Management Techniques.* 3rd Ed. Wildlife Soc. Ann Arbor.

TAYLOR, L. R. (1961). Aggregation, variance and the mean. *Nature, Lond.*, **189**, 732–735.

TAYLOR, L. R. (1963). Analysis of the effect of temperature on insects in flight. *J. Anim. Ecol.*, **32**, 99–112.

THOMAS, A. S. (1960). Changes in vegetation since the advent of myxomatosis. *J. Ecol.*, **48**, 287–306.

THOMAS, A. S. (1963). Further changes in vegetation since the advent of myxomatosis. *J. Ecol.*, **51**, 151–183.

THOMPSON, H. R. (1956). Distribution of distance to *n*th nearest neighbour in a population of randomly distributed individuals. *Ecology*, **37**, 391–394.

THORSTEINSON, A. J. (1960). Host selection in phytophagous insects. *A. Rev. Ent.*, **5**, 193–218.

TINBERGEN, L. (1960). The natural control of insects in pinewoods. 1. Factors influencing the intensity of predation by songbirds. *Archs neerl. Zool.*, **13**, 265–343.

USHER, M. B. (1975). Analysis of pattern in real and artificial plant populations. *J. Ecol.*, **63**, 569–586.

VARLEY, G. C., GRADWELL, G. R. and HASSELL, M. P. (1973). *Insect Population Ecology—an analytical approach.* Blackwell, Oxford.

WALOFF, N. and BLACKITH, R. E. (1962). The growth and distribution of the mounds of *Lasius flavus* (Fabricius) (Hym: Formicidae) in Silwood Park, Berkshire. *J. Anim. Ecol.*, **31**, 421–437.

WATSON, M., HULL, R., HAMLYN, B. and BLENCOWE, J. W. (1951). The spread of beet yellows and beet mosaic viruses in the sugar beet crop. I. Field observations on the virus diseases of sugar beet and their vectors, *Myzus persicae* Sulz., and *Aphis fabae*, Koch. *Ann. appl. Biol.*, **38**, 743–758.

WATT, A. S. (1947). Pattern and process in the plant community. *J. Ecol.*, **35**, 1–22.

WATT, A. S. (1955). Bracken versus heather, a study in plant sociology. *J. Ecol.*, **43**, 490–506.

WAY, M. J. and BANKS, C. J. (1967). Intra-specific mechanisms in relation to the natural regulation of numbers of *Aphis fabae* Scop. *Ann. appl. Biol.*, **59**, 189–205.

WELCH, J. R. (1960). Observations on deciduous woodland in the eastern province of Tanganyika. *J. Ecol.*, **48**, 557–573.

WENT, F. W. (1942). The dependence of certain annual plants on shrubs in southern California deserts. *Bull. Torrey Bot. Club*, **69**, 100–114.

WESTOBY, M. (1977). Self-thinning driven by leaf area, not by weight. *Nature*, **265**, 330–331.

WHITE, J. and HARPER, J. L. (1970). Correlated changes in plant size and number in plant populations. *J. Ecol.*, **58**, 467–485.

WHITTAKER, R. H. and FEENY, P. P. (1971). Allelochemics: chemical interactions between species. *Science 171*, 757–770.

WILHM, J. (1972). Graphic and mathematical analyses of biotic communities in polluted streams. *Ann. Rev. Ent.*, **17**, 223–251.

WILLIAMS, C. B. (1940). The analysis of four years' captures of insects in a light trap. Part 2. The effect of weather conditions on insect activity;

and the estimation and forecasting of changes in the insect population. *Trans. R. Ent. Soc. Lond.*, **90**, 227–306.

WILLIAMS, C. B. (1958). *Insect Migration.* Collins, London.

WILLIAMS, C. B. (1961). Studies on the effect of weather conditions on the activity and abundance of insect populations. *Phil. Trans. B.*, **244**, 331–378.

WILLIAMS, C. B. (1964). *Patterns in the Balance of Nature and Related Problems in Statistical Ecology.* Academic Press, London.

WILLIAMS, J. T. and VARLEY, Y. W. (1967). Phytosociological studies of some British grasslands. *Vegetatio*, **15**, 169–189.

WILLIAMS, N. V. and DUSSART, G. B. J. (1976). A field course survey of three English river systems. *J. Biol. Educ.*, **10**, 4–14.

WILLIAMS, W. T. and LAMBERT, J. M. (1959). Multivariate methods in plant ecology. I. Association-analysis in plant communities. *J. Ecol.*, **47**, 83–101.

WRATTEN, S. D. (1974). Aggregation in the birch aphid, *Euceraphis punctipennis* (Zett.) in relation to food quality. *J. Anim. Ecol.*, **43**, 191–198.

YODA, K., KIRA, T., OGAWA, H. and HOZUMI, H. (1963). Self-thinning in overcrowded pure stands under cultivated and natural conditions. *J. Biol. Osaka Cy. Univ.*, **14**, 107–129.

Tables

Table 1. The probabilities associated with values of z in a normal distribution. (After Lindley, D. V. and Miller, J. C. P. (1953). *Cambridge Elementary Statistical Tables.* Cambridge University Press.)

z	p_1	p_2
0·0	0·500	1·000
0·1	0·460	0·920
0·2	0·421	0·841
0·3	0·382	0·764
0·4	0·345	0·689
0·5	0·309	0·617
0·6	0·274	0·549
0·7	0·242	0·484
0·8	0·212	0·424
0·9	0·184	0·368
1·0	0·159	0·317
1·1	0·136	0·271
1·2	0·115	0·230
1·3	0·097	0·193
1·4	0·081	0·162
1·5	0·067	0·134
1·6	0·055	0·110
1·7	0·045	0·089
1·8	0·036	0·072
1·9	0·029	0·057
*1·96	0·025	0·050
2·0	0·023	0·046
2·1	0·018	0·036
2·2	0·014	0·028
2·3	0·011	0·021
2·4	0·008	0·016
2·5	0·006	0·012
*2·58	0·005	0·010
2·6	0·005	0·009
2·7	0·004	0·007
2·8	0·003	0·005
2·9	0·002	0·004
3·0	0·001	0·003
3·1	0·001	0·002
3·2	0·001	0·001
3·3	0·001	0·001
3·4	0·000	0·001
3·5	0·000	0·000

p_1 = the probability of a value being more extreme than z (a one-tailed probability).
p_2 = the probability of a value being more extreme than either $+z$ or $-z$ (a two-tailed probability).
*Critical values of z corresponding to the 0·05 and 0·01 levels in the two-tailed test have been given to two decimal places.

Tables 2A and 2B. Critical values for r in the runs test for n_1 or n_2 from 2 to 20. Any calculated value of r equal to or smaller than the values given in Table 2A or equal to or larger than those given in Table 2B is significant at the 0.05 level. (After Swed, F. S. and Eisenhart, C. (1943). Tables for testing Randomness of Grouping in a sequence of Alternatives. *Ann. Math. Statist.* Vol. **14,** 83–86. From Siegel, S. (1956). *Nonparametric Statistics for the Behavioural Sciences.* McGraw-Hill.)

2A

n_1 \ n_2	2	3	4	5	6	7	8	9	10	11	12	13	14	15	16	17	18	19	20
2											2	2	2	2	2	2	2	2	2
3					2	2	2	2	2	2	2	2	2	3	3	3	3	3	3
4				2	2	2	3	3	3	3	3	3	3	3	4	4	4	4	4
5			2	2	3	3	3	3	3	4	4	4	4	4	4	4	5	5	5
6		2	2	3	3	3	3	4	4	4	4	5	5	5	5	5	5	6	6
7		2	2	3	3	3	4	4	5	5	5	5	5	6	6	6	6	6	6
8		2	3	3	3	4	4	5	5	5	6	6	6	6	6	7	7	7	7
9		2	3	3	4	4	5	5	5	6	6	6	7	7	7	7	8	8	8
10		2	3	3	4	5	5	5	6	6	7	7	7	7	8	8	8	8	9
11		2	3	4	4	5	5	6	6	7	7	7	8	8	8	9	9	9	9
12	2	2	3	4	4	5	6	6	7	7	7	8	8	8	9	9	9	10	10
13	2	2	3	4	5	5	6	6	7	7	8	8	9	9	9	10	10	10	10
14	2	2	3	4	5	5	6	7	7	8	8	9	9	9	10	10	10	11	11
15	2	3	3	5	5	6	6	7	7	8	8	9	9	10	10	11	11	11	12
16	2	3	4	4	5	6	6	7	8	8	9	9	10	10	11	11	11	12	12
17	2	3	4	4	5	6	7	7	8	9	9	10	10	11	11	11	12	12	13
18	2	3	4	5	5	6	7	8	8	9	9	10	10	11	11	12	12	13	13
19	2	3	4	5	6	6	7	8	8	9	10	10	11	11	12	12	13	13	13
20	2	3	4	5	6	6	7	8	9	9	10	10	11	12	12	13	13	13	14

2B

n_1 \ n_2	2	3	4	5	6	7	8	9	10	11	12	13	14	15	16	17	18	19	20
2																			
3																			
4				9	9														
5			9	10	10	11	11												
6			9	10	11	12	12	13	13	13	13								
7				11	12	13	13	14	14	14	14	15	15	15					
8				11	12	13	14	14	15	15	16	16	16	16	17	17	17	17	17
9					13	14	14	15	16	16	16	17	17	18	18	18	18	18	18
10					13	14	15	16	16	17	17	18	18	18	19	19	19	20	20
11					13	14	15	16	17	17	18	19	19	19	20	20	20	21	21
12					13	14	16	16	17	18	19	19	20	20	21	21	21	22	22
13						15	16	17	18	19	19	20	20	21	21	22	22	23	23
14						15	16	17	18	19	20	20	21	22	22	23	23	23	24
15						15	16	18	18	19	20	21	22	22	23	23	24	24	25
16							17	18	19	20	21	22	22	23	23	24	25	25	25
17							17	18	19	20	21	22	23	23	24	25	25	26	26
18							17	18	19	20	21	22	23	24	25	25	26	26	27
19							17	18	20	21	22	23	23	24	25	26	26	27	27
20							17	18	20	21	22	23	24	25	25	26	27	27	28

Index